Shifting Shorelines

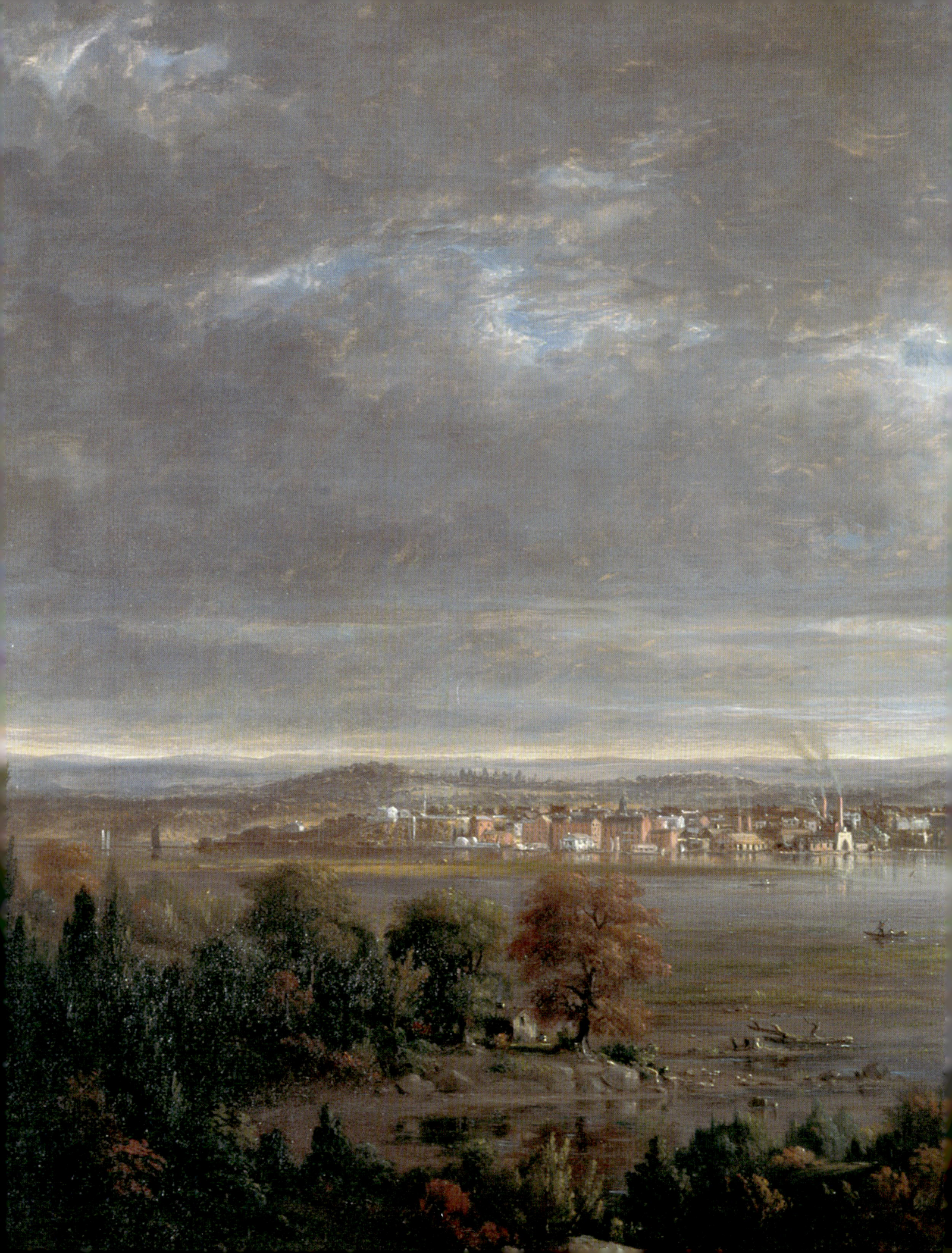

Shifting Shorelines
Art, Industry, and Ecology along the Hudson River

Annette Blaugrund, Betti-Sue Hertz,
Elizabeth Hutchinson, and Dorothy M. Peteet

with contributions by Ross Barrett, Tracy Brown, Victoria Horrocks,
May Joseph, Anthony Papa, Christopher Pickerell, Steven Schimmrich,
and Jonathan Weinberg

WALLACH ART GALLERY
Columbia University in the City of New York

CONTENTS

7 FOREWORD AND ACKNOWLEDGMENTS
Betti-Sue Hertz

11 LENDERS TO THE EXHIBITION

13 Introduction
Annette Blaugrund, Betti-Sue Hertz,
Elizabeth Hutchinson, Dorothy M. Peteet

17 From Relative to Resource
*Depicting Ecology and Industry
along the Hudson, 1650–1850*
Elizabeth Hutchinson

31 Early New York Oyster Jars
Christopher Pickerell

35 Landscape Painting and the
Hudson Valley Brick Industry
Ross Barrett

41 Omissions from the Canon
*Artists' Representations of Industry along
the Hudson River in the 19th and 20th Centuries*
Annette Blaugrund

59 Industrial Harmony
Connecting Industry along the Hudson River Railroad
Victoria Horrocks

63 **Hudson Marshes**
Archives of Biodiversity, Climate Change, and Human Impact
Dorothy M. Peteet

77 **The Geology of Shifting Shorelines**
Steven Schimmrich

81 **From Degradation to Restoration**
Shifting the Hudson's Storyline
Tracy Brown

85 **The Hudson River and Anthony Papa**
Anthony Papa

89 **Hudson River as Sink**
Waste, Pollution, and the Restoration of Nature
May Joseph

103 **On the Waterfront Again**
Jonathan Weinberg

107 **Extraction, Products, and the Environment**
Materiality in Contemporary Art of the Hudson River
Betti-Sue Hertz

121 ARTISTS' ROUNDTABLE
129 CATALOGUE OF THE EXHIBITION
136 ABOUT THE CONTRIBUTORS
141 HUDSON RIVER ADVOCACY RESOURCES
143 PHOTOGRAPHY CREDITS

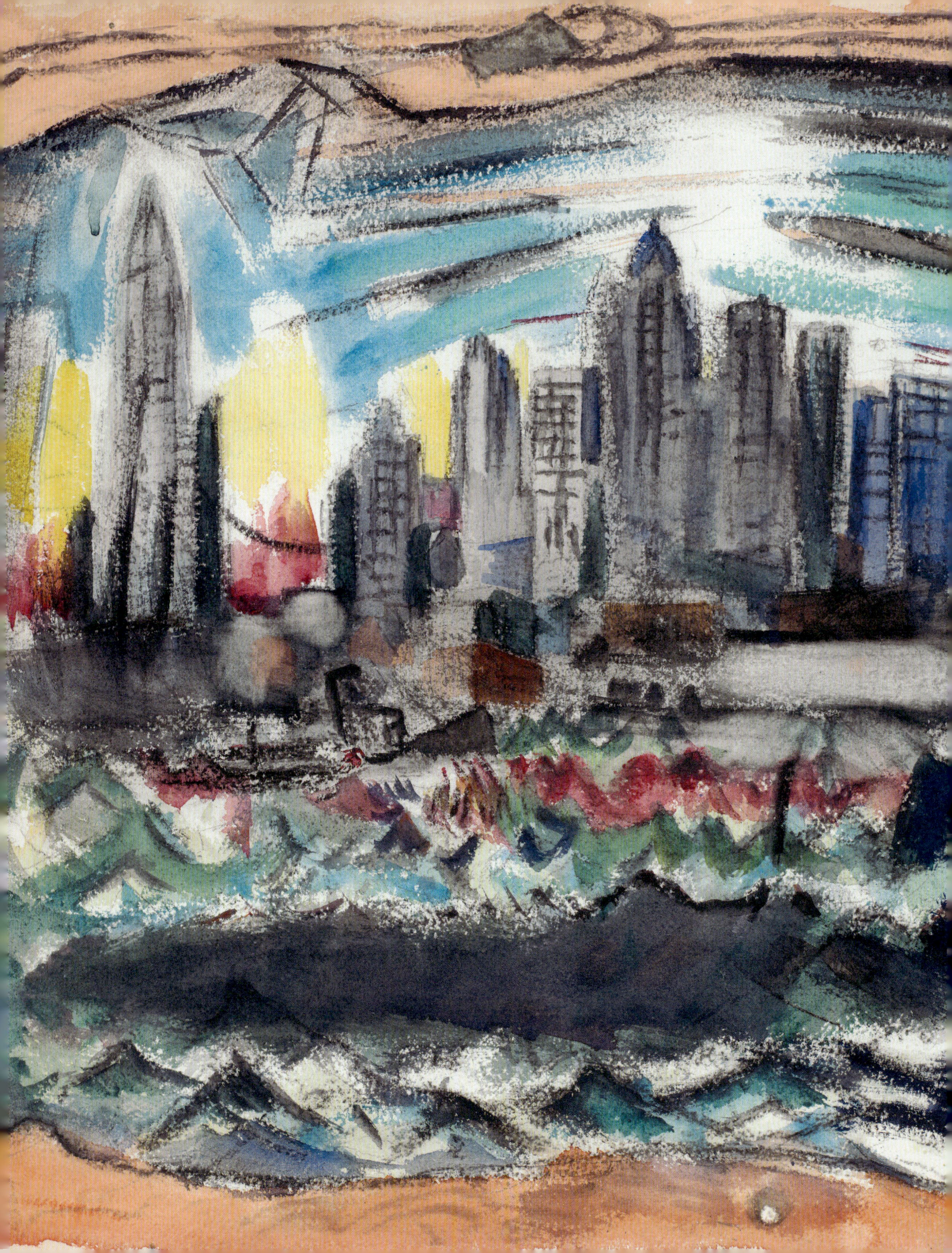

FOREWORD AND ACKNOWLEDGMENTS

Betti-Sue Hertz

Shifting Shorelines: Art, Industry, and Ecology Along the Hudson River features landscapes, ecology, shoreline industries, and the people who lived nearby and labored therein. The works on view, organized chronologically and by industry, tell the river's history through the experiences of historical and contemporary artists. Their engagement and, dare I say, love for the Hudson River, when it is sublime and inspiring and even when it is under duress, are preserved in the images and objects on display. Emphasized by the view of the river from the gallery, the dialogue among the works in the exhibition brings the urgency of a nuanced consideration of the Hudson's contributions—past, present, and future—to life along its shores.

When the Wallach Art Gallery moved into the Lenfest Center for the Arts in 2017, we gained a beautiful exhibition space with a south-facing wall overlooking the city. We are also graced with a view of the Hudson River and New Jersey seen through the urban landscape at the western edge of 125th Street. For thousands of years, the natural topography of this site formed a valley and a small, sheltered cove that was excellent for fishing and launching boats. The West Harlem Piers park, inaugurated in 2008, hugs the water, emphasizing this feature of the shoreline. The Henry Hudson Parkway and the railroad tracks rise above street level just east of the park. A celebrated elevated steel viaduct, built in 1901 and comprising twenty-six spans, partially blocks a full view of the river and park. The north and south approaches of the viaduct are faced with Mohawk Valley limestone with Maine granite trimmings, and the ornamental facework is coursed ashlar. Warehouses that once served local industries sit below and just west of the viaduct. They were storehouses for supplies and goods delivered by boats plying their way up and down the river and via railroad. These buildings now house restaurants and until recently Fairway supermarket. Some buildings are empty, waiting for redevelopment by Columbia University. Two blocks east of the river on the southside of 125th across from the Lenfest Center is Prentiss Hall, home to Columbia's graduate visual arts studios located in the former Sheffield Farms milk factory built in 1911. This structure, which retains its historical facade, is also emblematic of multifaceted history of the neighborhood.

I want to wholeheartedly thank the many people who have made this exhibition and publication possible, especially the co-curators and

John Marin. *Lower Manhattan from River, No. 1* (detail), p. 56

co-editors of this volume, Annette Blaugrund, Elizabeth Hutchinson, and Dorothy M. Peteet, fellow travelers on the long road of conceiving, developing, and manifesting a new vision of the art history of the river. The germ of the idea for the exhibition came about in a conversation with Annette, who, with her passion for, and expertise on, the art of the Hudson River, proposed that we work on a project together. We invited two Columbia colleagues to join us: the art historian Hutchinson, who specializes in American art; and the scientist Peteet, who studies the river's marshes. With this incredible talent, we began to research our subject and select objects that could demonstrate industry and its interface with nature.

There are so many moving parts to an exhibition of this scale and ambition. I want to thank our student curatorial researchers, Christine Yi-Ting Chen and Victoria Horrocks, both of whom have been invaluable in helping with the coordination of the curatorial team, research, and record-keeping.

In November 2021 we held a roundtable via Zoom to speak with scholars and artists knowledgeable about the representation and the history of the Hudson. I am grateful to Myra Young Armstead, Meredith David, May Joseph, Karl Kusserow, Kate Menconeri, Eric K. Washington, and Kayla Weisdorf for their willingness to participate and share their wisdom.

We are grateful to the following scholars and advocates in the arts and environmental spheres who helped us better understand industry and its impact on the environment: Yona Backer, Joe Baker, Clara Chang, Hadrien Coumans, Lauren Daisley, Sofia Thieu D'Amico, Joseph Diamond, Margi Hofer, Wendy Ikemoto, Jennifer Krieger, Marilyn Kushner, Jonathan Lothrop, Jessica MacLean, Mark Mitchell, Todd Ommen, Christopher Pickerell, Erin Riley, Matthew Sanger, Gwendolyn Saul, Peggy Shepard, Nancy Siegel, Amanda Sutphin, Monique Tynall, Jo-Ann Wong, William Zeisel, and Curtis Zunigha.

We could not have put together the exhibition without the support of the staff at other institutions who shepherded our inquiries and loan requests through the various channels for approval. Thanks go to the following directors, curators, acquisition committee members, gallerists, and registrars for partnering with us—we could not have achieved our goal of presenting diverse works without their support and guidance: L. Lynne Addison, Thomas Baldwin, Chris Browser, Pamela Bransford, Rebecca Chartier-Hobson, Karen Convertino, Matthew Coolidge, Alice N. Culclasure, Emily Cushman, Carli DeFillo, Alyssa Dreliszak, Tricia Earl, Tara Emsley, Emily Foss, Kathleen A. Foster, Alexander Garcia, Claire Elise Glenn, Mason Klein, Marissa Klein-Kundrath, Elizabeth Largi, Emily Lenz, Jennifer Levy, Jonathan Lothrop, Greg Lulay, Julie Maguire, Crawford Mann, Karl McCool,

Sheryl McMahan, Barbara McNulty, Shadi Mirsepassi, Caitlin Monaco, Janet Moore, Antonia Moser, Antonio Murzi, Kristi Reese, Gwendolyn Saul, Mark Schlemmer, Aurora Schmit, Patrick Shepler, Diane Shewchuk, Charles Siskin, Michael Somple, Anthony Speiser, Barbi Spieler, Sarah Tignor, Vanessa D. Thaxton, Monique Tyndall, Erika Umali, Cassie Ward, Deedee Wigmore, Sarah C. Wolfe, and Andrea Zlotowitz.

The contributors to this catalogue taught us that there are many perspectives to consider when mounting an exhibition of this scope and scale. The knowledge of these talented and engaged colleagues has greatly enhanced this publication. Essays by the four curators are joined by texts by Ross Barrett, Tracy Brown, Victoria Horrocks, May Joseph, Anthony Papa, Christopher Pickerell, Steven Schimmrich, and Jonathan Weinberg. The copyediting expertise of Pamela Barr and superb design by Laura Lindgren have been integral to the realization of this beautiful publication.

Discussions with the living artists represented in the show led to an understanding of how environmental concerns are ever present against the backdrop of complex histories in their thinking, perspectives, and work. I thank Shi Guorui, David Hammons, Every Ocean Hughes, An-My Lê, Alan Michelson, Anthony Papa, and Lisa Sanditz, for their vision, honesty, and willingness to lend their work to this exhibition. Athena LaTocha, Courtney M. Leonard, Marie Lorenz, and Jean-Marc Superville Sovak participated in the artists' roundtable, which is published here in condensed form.

Everything that happens at the Wallach Art Gallery is supported by the team that works tirelessly to ensure that every exhibition is produced at the highest level possible. Every day they strive to surpass the goals we set for ourselves and activate the gallery's mission. I would like to thank Eddie Bartolomei, Nathaly Berrio-Diaz, Evan Clinton, Stephanie Litchfield, Daniel Lopez, Jennifer Mock, Andrew Page, Jeanette Silverthorne, and Zachary Valdez. For *Shifting Shorelines*, we were fortunate to have Dan Kershaw and Laura Lindgren join us as consulting exhibition designers.

The realization of *Shifting Shorelines* has been an especially gratifying experience for me, and I believe that as a presentation space in a leading university, nestled on the shoreline of the Hudson, we are uniquely positioned to contribute to the art history of the region through a transhistorical interdisciplinary view. I am grateful to the many people who supported this effort. I thank the current members of the Wallach Art Gallery Steering Committee—Diane Bodart, Matthew Buckingham, Roberto Ferrari, Aidan Ford Chisholm, Holger Klein, Jonathan Reynolds, Tomas Vu-Daniel, and Yasmine Yakuppur—and our advisory council members Winnie Feng, Jeffrey Hoffeld, Marley Blue Lewis, and Alan Wang for sharing their expertise and perspectives on the Wallach's endeavors.

We work closely with Elisa Heikkilä and the Arts and Sciences team in the Office of Alumni and Development on identifying partners to support our programs. We are thrilled that the trustees, directors, and staff at the Terra Foundation, Wyeth Foundation, Dr. Lee MacCormick Edwards Charitable Foundation, and Lunder Foundation—Peter and Paula Lunder Family have joined us in this endeavor, providing critically important funds. A gift from an anonymous donor, made in memory of Stanley Blaugrund, M.D., is also much appreciated.

I am indebted to Columbia University for its unwavering support for the Wallach Art Gallery and our goal of presenting important art and ideas to the public. I especially thank Amy Hungerford, Executive Vice President of Arts and Sciences, for providing the guidance and space to manifest a vision for the Wallach Art Gallery that meshes with Columbia's larger mission and allows for the freedom to shape a wide-ranging exhibition program inclusive of local and global artistic expressions.

LENDERS TO THE EXHIBITION

Albany Institute of History & Art
Estate of Alvin Baltrop, courtesy of Galerie Buchholz, New York
Brooklyn Museum
Center for Land Use Interpretation, Culver City, California
Cooper Hewitt, Smithsonian Design Museum, New York
Dallas Museum of Art
Electronic Arts Intermix, New York
Everson Museum of Art, Syracuse
Shi Guorui
Hampton University Museum, Virginia
Historic Hudson Valley, Pocantico Hills, New York
Historic Richmond Town, Staten Island
Donna Hogerhuis
Hudson River Museum, Yonkers
Every Ocean Hughes
Estate of Yvonne Jacquette, courtesy Mary Ryan Gallery, New York
The Jewish Museum, New York
The Johnson Collection, Spartanburg, South Carolina
An-My Lê, courtesy Marion Goodman Gallery, New York
Lebanon Valley College Fine Art Collection, Annville, Pennsylvania
Courtney M. Leonard
Library of Congress, Washington, D.C.
Marie Lorenz
The Metropolitan Museum of Art, New York
Alan Michelson
Montgomery Museum of Fine Arts, Alabama
Munson-Williams-Proctor-Arts-Institute, Utica

Antonio Mursi and Diana Morgan Collection
Museum of the City of New York
Museum of Fine Arts, Boston
Museum of Modern Art, New York
Newington-Cropsey Foundation
New York City Archaeological Repository
New-York Historical Society
The New York Public Library, Miriam and Ira D. Wallach Division of Art, Prints, and Photographs
New York State Museum, Albany
Anthony Papa
Dorothy M. Peteet
Chris Pickerell
Philadelphia Museum of Art
Putnam History Museum, Cold Spring, New York
RISD Museum, Providence
Lisa Sanditz, courtesy Huxley Parlour Gallery, London
Staten Island Historical Society
Stockbridge-Munsee Band of Mohicans
Jean-Marc Superville Sovak
Telfair Museums, Savannah, Georgia
Washington County Museum of Fine Arts, Hagerstown, Maryland
Weatherspoon Art Museum, Greensboro, North Carolina
Whitney Museum of American Art, New York
Wichita Art Museum
D. Wigmore Fine Art, Inc., New York
Yale University Art Gallery, New Haven

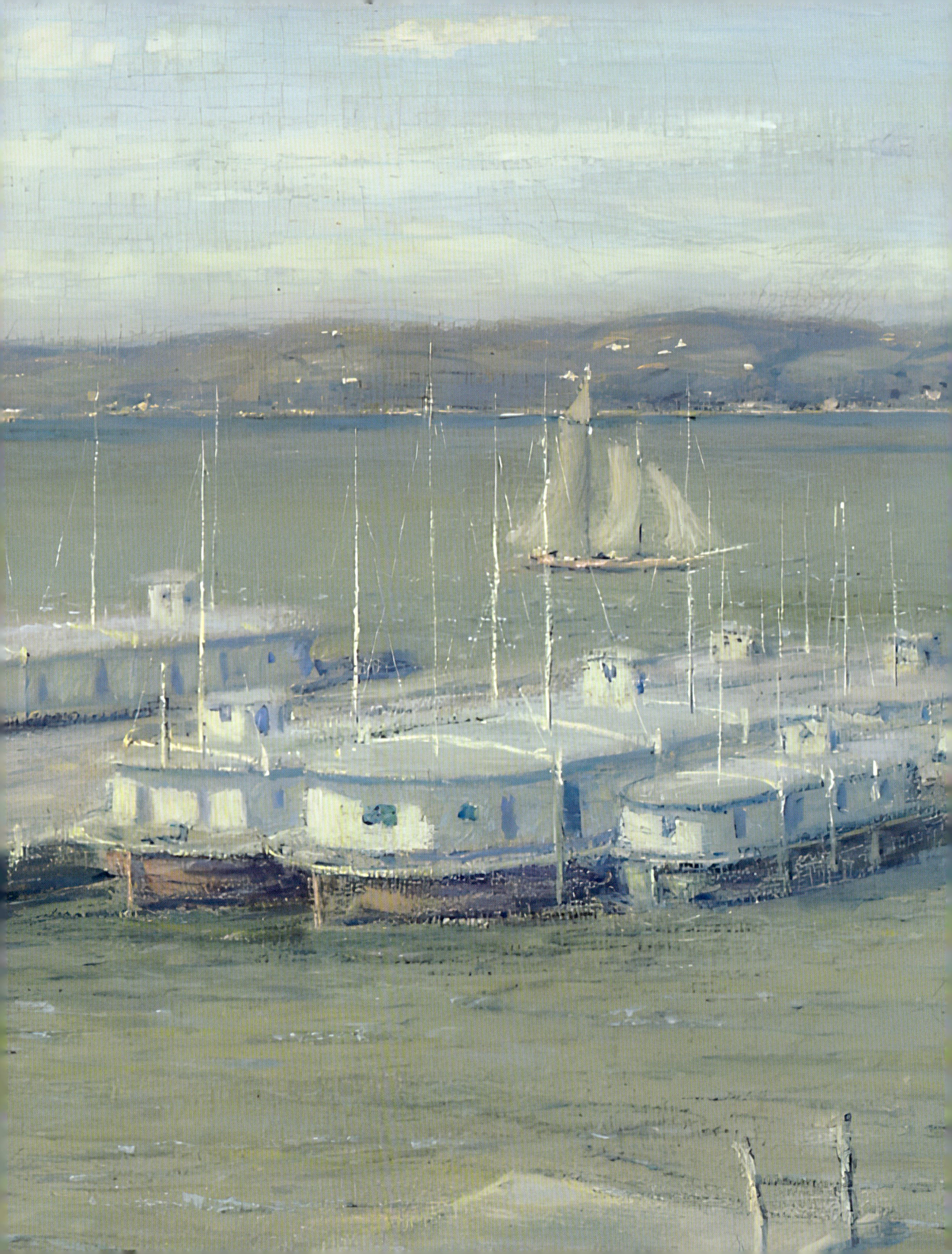

Introduction

Annette Blaugrund, Betti-Sue Hertz,
Elizabeth Hutchinson, Dorothy M. Peteet

In 2021 a team of four curators—each having different skills, academic interests, and experiences—convened to create an exhibition that would add a new dimension to the history of art and visual culture. Expanding on previous exhibitions and catalogues, we planned to trace how various industries impacted the Hudson River and its shoreline to reiterate concerns about the reclamation of the river and its environs to a wide museum audience. *Shifting Shorelines*—which includes paintings, prints, sculpture, film, photography, material culture, and scientific documentation—presents a counterpoint to the idealized landscapes by nineteenth-century Hudson River School painters that erased, obscured, and distanced the complex realities of the river. By the twentieth century, several artists changed course and began to foreground the industries that increasingly dotted the shoreline. In the twenty-first century, environmentally sensitive artists focused on exposing the damage that had become a conspicuous aspect of American modernization and industrialization.

Our research revealed how the growth of industry, unequal structures of ownership and labor, and manufactured products impacted the land, water, and air. The goal of challenging the early myth of the Hudson as a virgin wilderness prompted us to include Indigenous work from before European colonialism that presents a sustainable model for harnessing the region's resources. The scientist on our curatorial team contributed her expertise about the palaeoecological history and changes to the river as shown in selected works and evidenced in sediment core samples recording how human and industrial intrusion changed nature in ways that continue to affect the ecology. Contemporary artists and community consultants helped us understand current responses to the deterioration and reclamation of the shoreline.

Shifting Shorelines provides artistic records of the extraction of materials from the Hudson River environment and their manufacture by Indigenous, enslaved, and immigrant workers. Landscape views capture the transportation of goods to market and record the transformations wrought on the shoreline by the creation and abandonment of piers, plantations, factories, prisons, and railroads. Seen in this context, paintings that might initially seem to have a purely aesthetic purpose, such as Henry Golden Dearth's *Ice*

Henry Golden Dearth. *Ice Boats on the Hudson* (detail), p. 14

Fig. 1. Henry Golden Dearth (1864–1918). *Ice Boats on the Hudson*, ca. 1888–98. Oil on canvas, 18 × 29⅛ in. (45.7 × 74.0 cm). Telfair Museum of Art, Savannah, Georgia, Gift of Gari Melchers (1908.6)

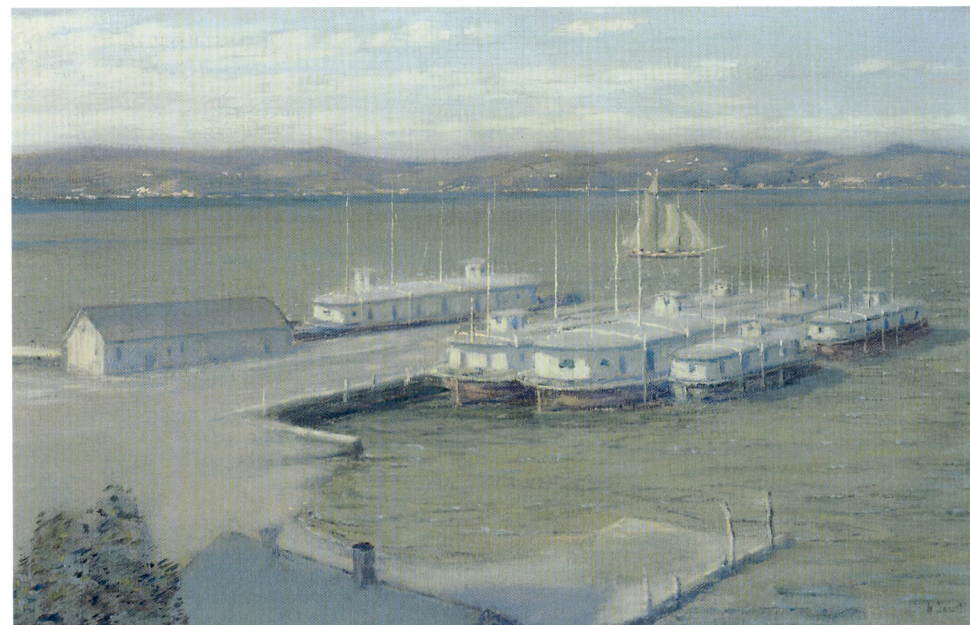

Boats on the Hudson (fig. 1) and Marie-François-Régis Gignoux's *Haverstraw on the Hudson* (fig. 18), become documents of forgotten industries whose formal and figurative dimensions both obscure and reveal the spread of the exploitation of the Hudson's natural resources. Even works that seem to emphasize the environment trace the spread of deforestation, the draining of wetlands, and the proliferation of railroads that cut off humans and animals from the waterfront. If nineteenth-century painters sought nature along the Hudson, early twentieth-century artists such as Abraham Leon Kroll and George Bellows presented industry triumphant. Yet, their works also betray the dangerous conditions that impacted workers and environment alike. Contemporary artists engage the afterlives of these histories through documentary and conceptual means.

The expanse of the project is emblemized by its historical scope from archaeological cultural objects through newly commissioned works and the desire to highlight the changing landscape—from early industries sited where the river flows into the bay in lower Manhattan, to small industries in the Bronx and towns in the Hudson Valley, to the logging and quarrying at the tidal limit of the river as an estuary in Troy. Throughout the exhibition nature and culture intersect with time and space to yield an ecological history. The tension between the health of the river and the productivity of its resources reveals itself in geographic fragments, through depictions of places such as Christopher Street, Inwood, Yonkers, Haverstraw, West Point, and Tivoli, each contributing to the larger visual narrative. Another perspective is offered by moving-image artists who navigate the river with scenic events of light

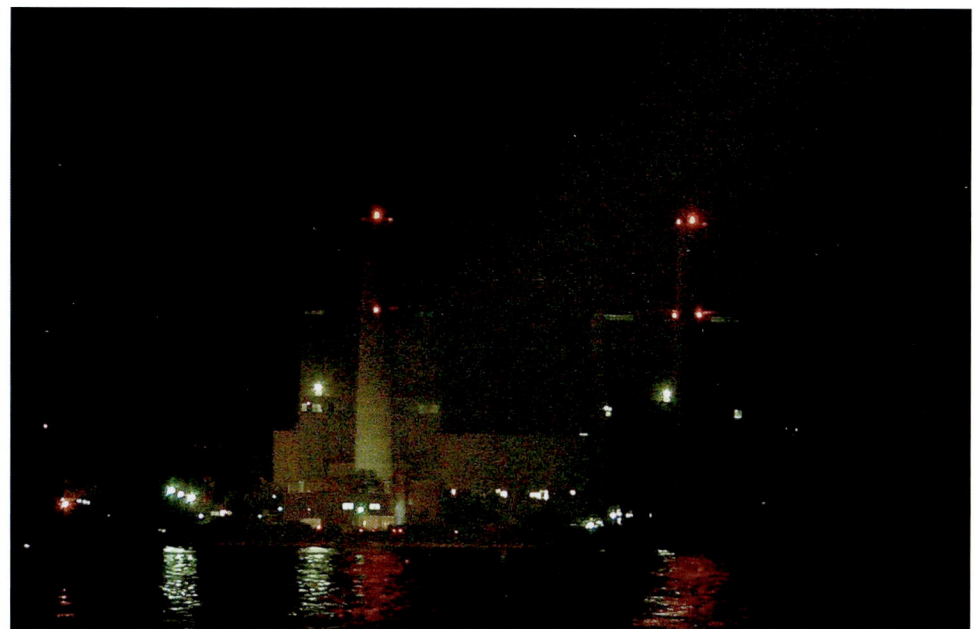

Fig. 2. Alan Michelson (b. 1953). *Shattemuc*, 2009, video still. HD video, stereo soundtrack with original music by Laura Ortman, 31 min. Collection of the artist

and heavy industry, farming, boating, recreation, building projects, and the intentionally changed shorelines, impacted habitats, forests, and marshes. The patchwork of developed, manipulated, and undeveloped landscapes is experienced through views from a boat or a plane. To represent the river as a coherent linear whole, the Center for Land Use Interpretation's *Up River: Man-Made Sites of Interest on the Hudson from the Battery to Troy*) (fig. 48), a collection of 84 sequential images spanning 130 miles, presents the shoreline with sweeping views from above—factories, avenues, prisons, power plants, quarries, parks, condos, ruins, and redevelopments—with special attention to the efforts of environmentalists. Its northernmost image is the Federal Dam in Troy. Alan Michelson's video *Shattemuc* (fig. 2), shot at night from a former police boat equipped with a searchlight aimed at the shoreline, retraces part of Henry Hudson's voyage up the river. In dimly lit views it collapses the bloody history of contact with natural beauty and the environmental hazards of industrial and real estate development. The prevalence of the panoramic view in the nineteenth century influenced the iconic Hudson River School painters. These contemporary artists revisit the methodology of the overview, not as a vehicle for picturing the ideal but as a haunting record of the past, juxtaposed with the Hudson's recent realities.

Retracing the arc of this history has deepened our understanding of the conflicts between production and environmentalism and taught us that repair is central to overcoming this dynamic and troubling history. Our aim is to bring attention to distinct periods in that historical trajectory and to signal an urgency for remediating the damage recorded in the works in the exhibition.

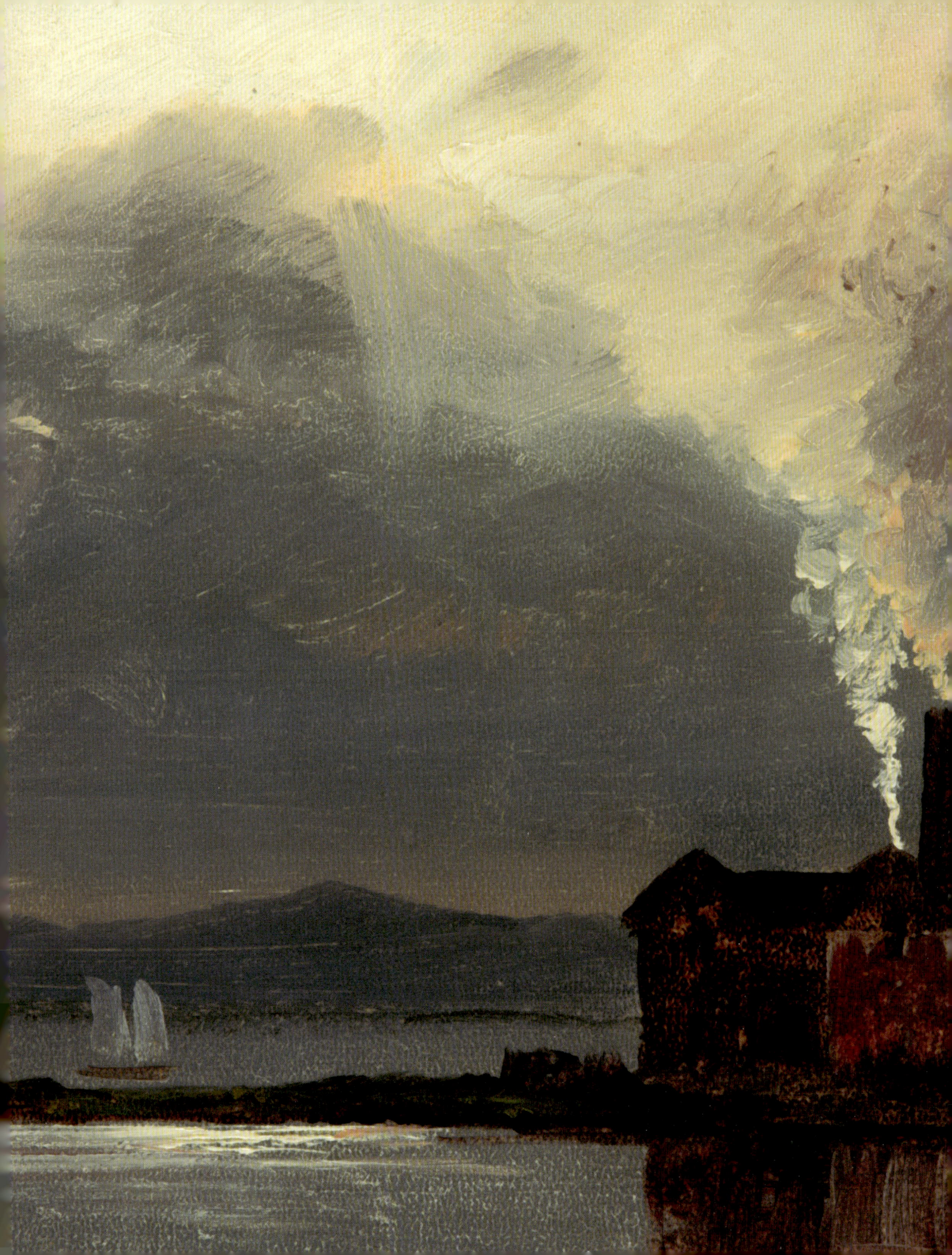

From Relative to Resource

Depicting Ecology and Industry along the Hudson, 1650–1850

Elizabeth Hutchinson

> It is a country well-adapted for our people ... since it seems to lack nothing needed for the subsistence of man.
> —Johannes de Laet, 1625[1]

When Europeans first saw the Hudson River, they marveled at the abundance of resources to be had along a single, easily navigable route. Henry Hudson recorded the availability of timber for shipbuilding and making casks, as well as the abundance of fish, and speculated about how to develop the copper and iron he observed being used by Indigenous people.[2] The flourishing Hudson River environment was a result of geological history, which over millennia laid down veins of metal; the last Ice Age, which deposited sediments rich with clay and minerals; and centuries of Native cultivation of the land. This history created the conditions that are celebrated in the earliest works of art in this exhibition, objects and images that trace an important shift in the relationship between the residents of the Hudson shorelines to the environment. The artwork produced by the Indigenous peoples of the region demonstrates their reciprocal relationship with the plants and animals that sustained them. European artists captured a landscape whose riches seemed inexhaustible, depicting the river and its environs as a resource with little regard for conservation.

Early dwellers along the Hudson, the Lenape along the southern stretch of the river and the Mohicans along the upper Hudson, practiced seasonal subsistence activities that took advantage of the riparian and coastal environments nourished by the tidal estuary. This habitat attracted members of other regional tribes for hunting, fishing, harvesting, and trade. The shores of the Hudson are populated with numerous middens that hold records of Lenape and Mohican life, including shells, netsinkers for fishing, pottery fragments, arrowheads, and the remains of plants, fish, and game. Abundant oyster shells reference summer feasts by small groups traveling along the Shattemuc—Lenape for "river that flows both ways"; Muhheakantuck, in Mohican—for fishing, hunting, gathering, and trade over countless generations. The oldest known shell midden has been dated to roughly 6950

Frederic Edwin Church. *Hudson River with Factory by Moonlight* (detail), p. 25

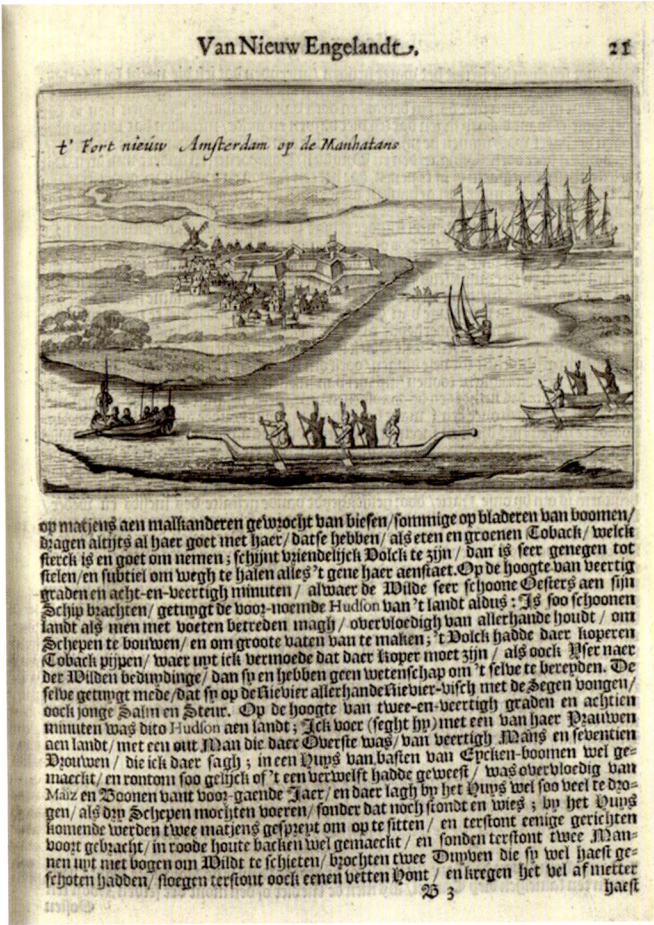

Fig. 3. Kryn Frederycks (17th century). *T'Fort Nieuw Amsterdam op de Manhatans*, 1651. Engraving, 3¼ × 4⅞ in. (8.2 × 12.4 cm). Isaac Newton Phelps Stokes Collection, The Miriam and Ira D. Wallach Division of Art, Prints and Photographs, The New York Public Library, Astor, Lenox and Tilden Foundations

BCE, shortly after the salinity of the river became ideal for oysters.[3] Clams also lived in the brackish waters of the tidal estuary, and their shells were used to create tools, personal adornments, and wampum—small cylindrical beads that were strung together for ornamental and ceremonial use and, later, for exchange with settlers. Shell continues to be an important material for Indigenous cultural and artistic production, but European settlement imperiled shellfish populations along the Hudson. By the mid-seventeenth century, the Dutch were already so worried about overharvesting that they set limits on the times and places for oystering.[4] Oyster shells were also an important resource. Crushed, they could serve as pavement for roads. Burned down and ground, they produced lime that was used in cement and plaster for buildings such as Trinity Church.[5] Oystering continued to be an important local industry into the early twentieth century, one that, as Christopher Pickerell discusses in this catalogue, attracted African American entrepreneurs.

A print of New Amsterdam from a 1651 book describing the American colonies for Europeans shows both Lenape and European boats crowding the harbor (fig. 3). The composition calls attention to European settlement through the inclusion of a windmill and fortifications, but it demonstrates the dominance of marine activities over terrestrial ones at the time. Early Dutch settlers described diverse fish, including striped bass, cod, herring, and mackerel, and sea mammals such as porpoises, seals, and even whales in the river. They observed the availability of wood, nuts, berries, and Indigenous "three sisters" gardens growing beans, corn, and squash, an intentional planting of crops that not only complement one another in terms of nutrition but also work together to maintain the richness of the soil in which they are grown.

Upriver it was the animals living along the shores of the Hudson, especially deer, beaver, and otters, that spurred the development of the first commercial trade from the colony. The Dutch built Fort Orange (present-day Albany), a trading post for acquiring furs and skins from Native men, and Fort New Amsterdam at the mouth of the river to protect the shipping of these

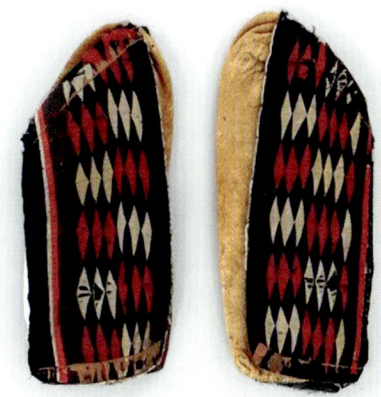 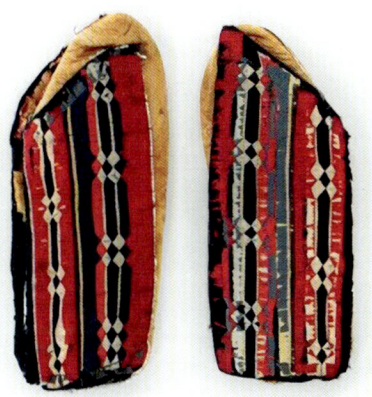

Fig. 4. Algonquian, Mohican. Moccasins, ca. 1800–1820. Deerskin embellished with ribbon and white beadwork, 9 × 4 in. (22.9 × 10.2 cm). New York State Museum, Albany (E-36295A-B)

goods to market. Initially, there was a balance of power between Native and non-native partners in the fur trade. Access to new goods inspired Indigenous women to play with color, design, and technique in the objects they made, as demonstrated by the Mohican works in this exhibition. A pair of deerskin moccasins acquired from Natives living in Rensselaer County about 1800 include long flaps embellished with cut silk ribbons and tiny white glass beads (fig. 4). The design and the decoration show the woman maker's knowledge. They are well constructed, and the embellishment suggests they were produced for a ceremonial occasion. The ribbon work, different on each flap, likely references important community relationships with visible and invisible forces in the Hudson River Valley—the plants that sustained community and the non-human beings that helped keep the world in balance.

Ultimately, the fur trade had devastating effects on Native communities up and down the river, introducing disease and spurring violence between tribes. The wealth accumulated through the trade led to increased European settlement, impeding access to vital resources and bringing war and forced dispossession of homelands. Artworks produced by Mohican makers in the late eighteenth and early nineteenth centuries provide clues to how communities sought ways to survive through this devastating history. While the moccasins show the perpetuation of cultural values and skills, basketry, developed for community use, was adapted for intercultural trade (fig. 30).[6] Gathering and preparing the materials for baskets from the wetlands along the Muhheakantuck reinforced relationships between people of different generations by offering the opportunity to pass on practices and stories. Well-constructed and practical plaited split ash storage baskets were popular with non-native neighbors and tourists, and weavers adapted their work to suit the needs of settlers—for example, by adding lids and producing items in a variety of sizes. The women makers

FROM RELATIVE TO RESOURCE

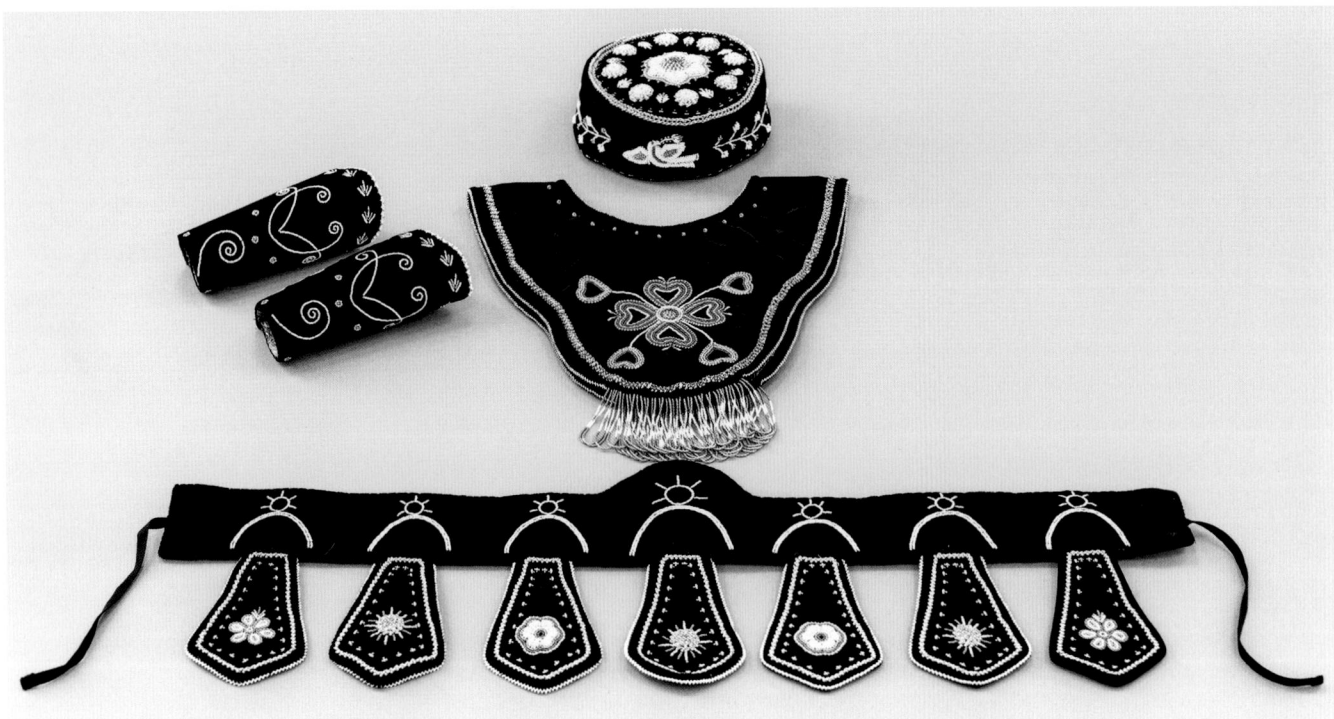

Fig. 5. Reva Fuhrman (b. 1985). Beaded dance regalia: hat, yoke, cuffs, and belt, 2018. Velvet and cotton fabric with glass beads, hat: 3¼ × 7½ in. (8.3 × 19.1 cm); yoke: 27½ × 15½ in. (68.9 × 39.4 cm); cuffs: 7¾ × 9¾ in. (19.7 × 24.8 cm); belt: 10½ × 34 in. (26.7 × 86.4 cm). New York State Museum, Albany (E-2018.39.1–4)

embellished the plaits with decorations made with potato stamps and indigo and other dyes. Contemporary Mohican artists continue these traditions, remaking community relationships with one another and with their homeland using materials and motifs that connect across generations. For example, Reva Fuhrman used purple and white beads that recall wampum to depict culturally significant Woodlands plants in the regalia she produced in 2018 (fig. 5). In 2021, 156 acres of land along the eastern shore of the Muckheantuck north of Castleton were returned to Stockbridge-Munsee Mohican Community.[7] The Lenape have not regained land along the river but retain a strong connection to Lenapehoking, their traditional territory.

Furs were not the only commodities exported from colonial New York. In 1773 Swedish naturalist Peter Kalm visited North America and observed, "New York sends many ships to the West Indies, with flour, corn, biscuit, timber, turns [casks], boards, flesh, fish, butter, and other provisions."[8] Several landowners replaced the forests along the Hudson, building saw- and gristmills to serve the needs of Caribbean plantations. For example, in the late seventeenth and early eighteenth centuries, the Netherlands-born Frederick Philipse and his heirs converted his 52,000-acre estate in what is now Westchester into farmland. Philipse owned several ships engaged in the slave trade and expanded his profits by also using them to send grain grown in New York to sugar plantations, including one owned by the family

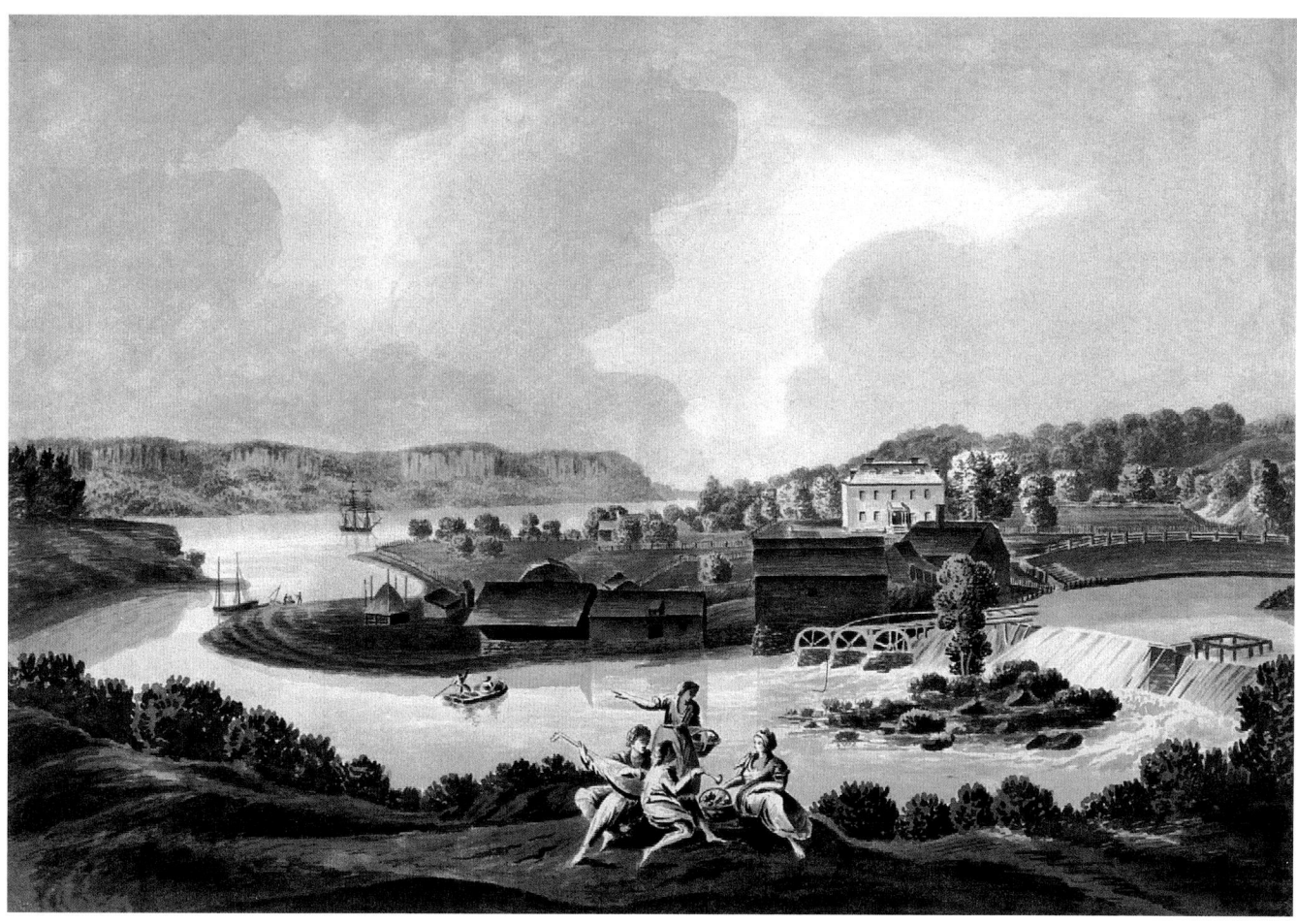

Fig. 6. American (18th century). *A View of Philipse Manor Hall*, 1784. Ink on paper, sight: 19¼ × 26¼ in. (48.9 × 66.7 cm). Historic Hudson Valley, Pocantico Hills, New York, Gift of La Duchesse de Talleyrand (PM.65.866)

in Barbados. The three gristmills in what are now Yonkers and Sleepy Hollow were among the first industrial installations along the Hudson. The 1784 drawing *A View of Philippse Manor Hall* shows the estate at the mouth of the Nepperhan River (fig. 6). While the land was cultivated by tenant farmers, the mill shown in the foreground, the cooperage (where barrels for shipping the milled grain were made), and the bakehouse (where ship's biscuits, a type of dry cracker with a long shelf life, were produced) were staffed by highly skilled enslaved men.[9] Enslaved labor was an essential part of the industrial development of the Hudson, from building the structures visible in the images shown here to producing bricks and other agricultural and industrial items in upstate factories.

The engraving of Philipsburg Manor reveals none of this complex history. Instead, the estate is seen at a picturesque distance that frames the sites of enslaved labor with nature, including the Palisades, and allegorical figures at leisure appear in the foreground to neutralize any industrial associations coming from the mill wheels behind them. As property of Loyalists,

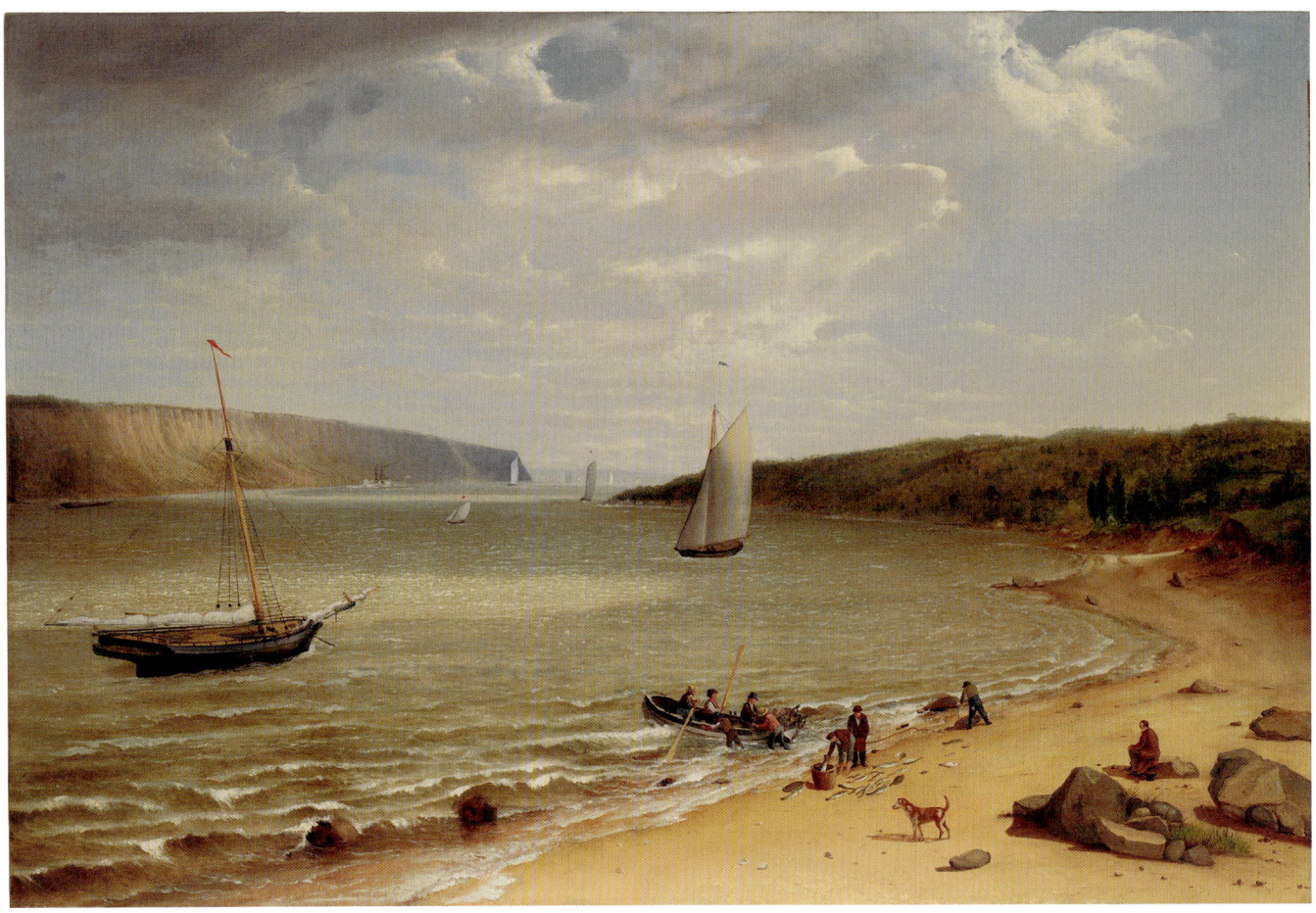

Fig. 7. Victor Gifford Audubon (1809–1860). *View of the Hudson River*, ca. 1845. Oil on canvas, 48 × 72 in. (121.9 × 182.9 cm). Museum of the City of New York, Gift of Miss Alice Lawrence (1938.188)

Philipsburg was seized and auctioned off after the Revolution. The gradual emancipation laws of 1799 and 1817 and challenges to the manor system by tenants led to the division and sale of other Hudson River estates, but by the early nineteenth century other industries were taking hold.

Blacks often continued to used the skills acquired during slavery, sometimes, as in the case of the African American potter Thomas Commeraw, producing oyster jars for shipment to plantations in the West Indies (fig. 11). These jars, made from clay harvested across the river in New Jersey, demonstrate his thoughtful craftsmanship—the openings were easy to seal with wax and cork, and their shape made them easy to ship. While these utilitarian jars did not receive the same colorful decoration that Commeraw applied to jars meant for domestic display, the salt glazes give them character and dignity that reinforces the message of their inscribed texts identifying the Black oystermen whose products were stored inside.

Most elite New York households held only a few enslaved persons. For example, artist Victor Gifford Audubon, son of the artist-naturalist John

James Audubon, relied on slave labor while growing up.[10] *View of the Hudson River* was painted on the Audubon estate in northern Manhattan not far from the Wallach Art Gallery (fig. 7). This ambitiously scaled canvas, executed during the years Audubon was seeking acceptance as an Associate of the National Academy of Design, depicts a distinctly Hudson River scene focused on Euro-American fishermen dropping off a basket of fish. The sloop, shown with its sails furled, was a typical vessel for moving cargo up and down the river even after Robert Fulton introduced steamboats in 1807, and several of them can be recognized by their single masts and shallow drafts in paintings in this exhibition. By including only one steamboat in the far background and filling the canvas with uncrowded waters, a largely empty beach, and an open sky, Audubon's view is somewhat nostalgic for 1845, when the river would have seen much more commercial traffic, including new iron-hulled steamboats such as the "Iron Witch" (cat. 20). But Audubon's painting is interesting for more than its conventionally picturesque rebalancing of the visual relationship between natural and man-made elements. It also demonstrates how an extractive relationship to nature had taken hold by Audubon's time. The underlying narrative describes the harvesting of fish from the river, first by being taken onto a large sloop, then transferred to an oar-powered dinghy, and then off-loaded onto the beach for local transport, in this case, likely to the Audubons' kitchen. The fish in this painting have been identified as striped bass whose populations, once high, were being diminished with the growth of commercial fishing.

While farming and fishing dominated the colonial economy, the extraction of industrial materials from the Hudson River environs began in the early eighteenth century. Shortly after Audubon completed his picture, construction of the Hudson River Railroad began, first connecting Troy with what is now Rensselaer and, in 1846, beginning an extension that reached New York City in 1851. The view we see in this painting became impossible once the tracks traced the shoreline.

Early industries attracted laborers to the region, and not only by their own volition. Sing Sing prison was established on 130 acres along the Hudson in 1826 and immediately received one hundred prisoners who promptly began to excavate marble from the local Tuckahoe quarry to build the first cellblock. The prisoners continued to work in stonecutting and also produced shoes, hats, wooden barrels, kitchen utensils, and other items for third-party contractors.[11] Convict labor was understood as more than punishment. It was believed that such efforts could lead criminals toward moral rehabilitation. Nineteenth-century Americans were proud of their carceral institutions and frequently visited them to observe their modern facilities and orderly operations.[12]

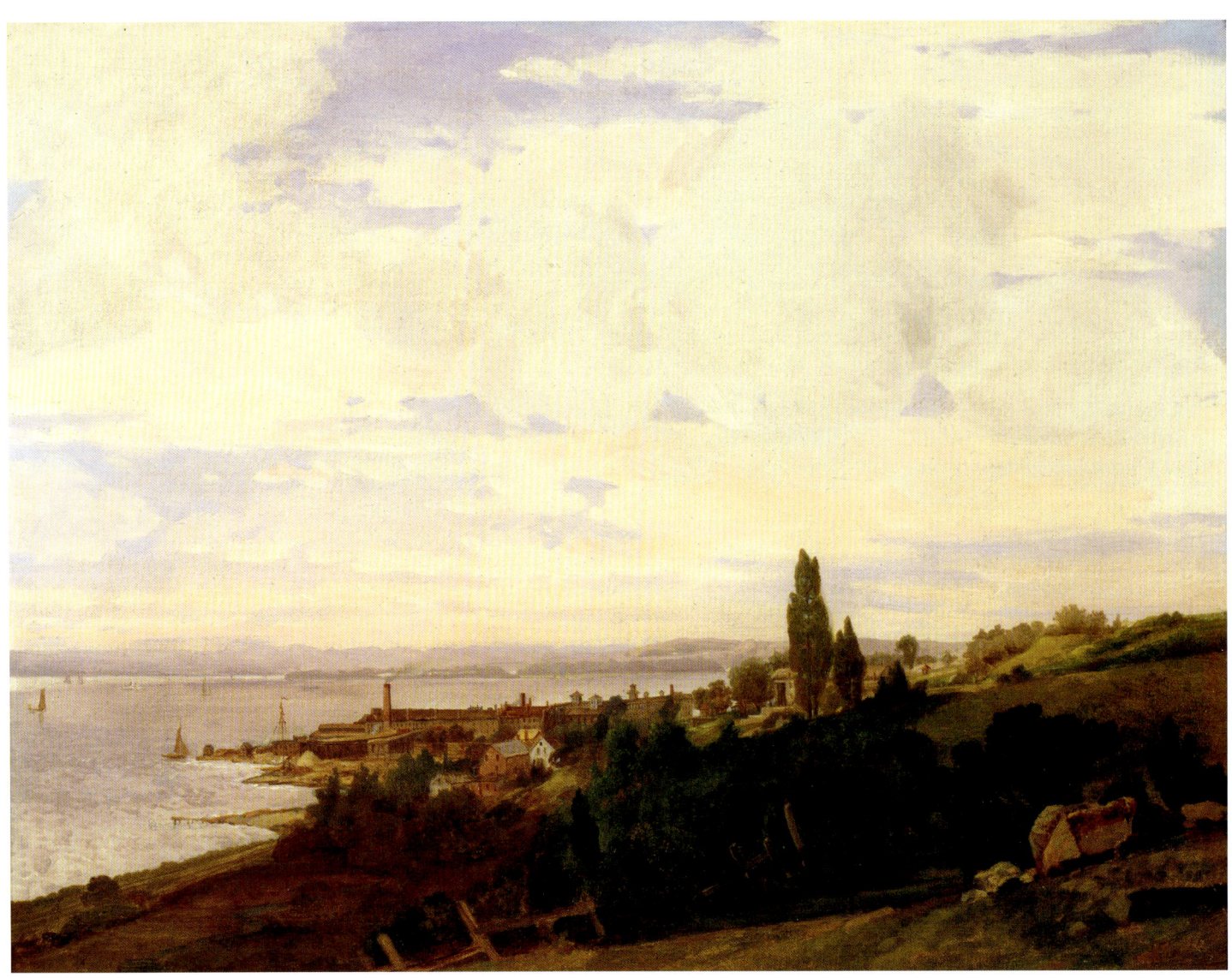

Fig. 8. Joseph Vollmering (1810–1887). *View on the Hudson Near Sing Sing, New York*, 1845–50. Oil on canvas, 21½ × 28 in. (54.6 × 71.1 cm). New-York Historical Society, Purchase, Watson Fund (1978.57)

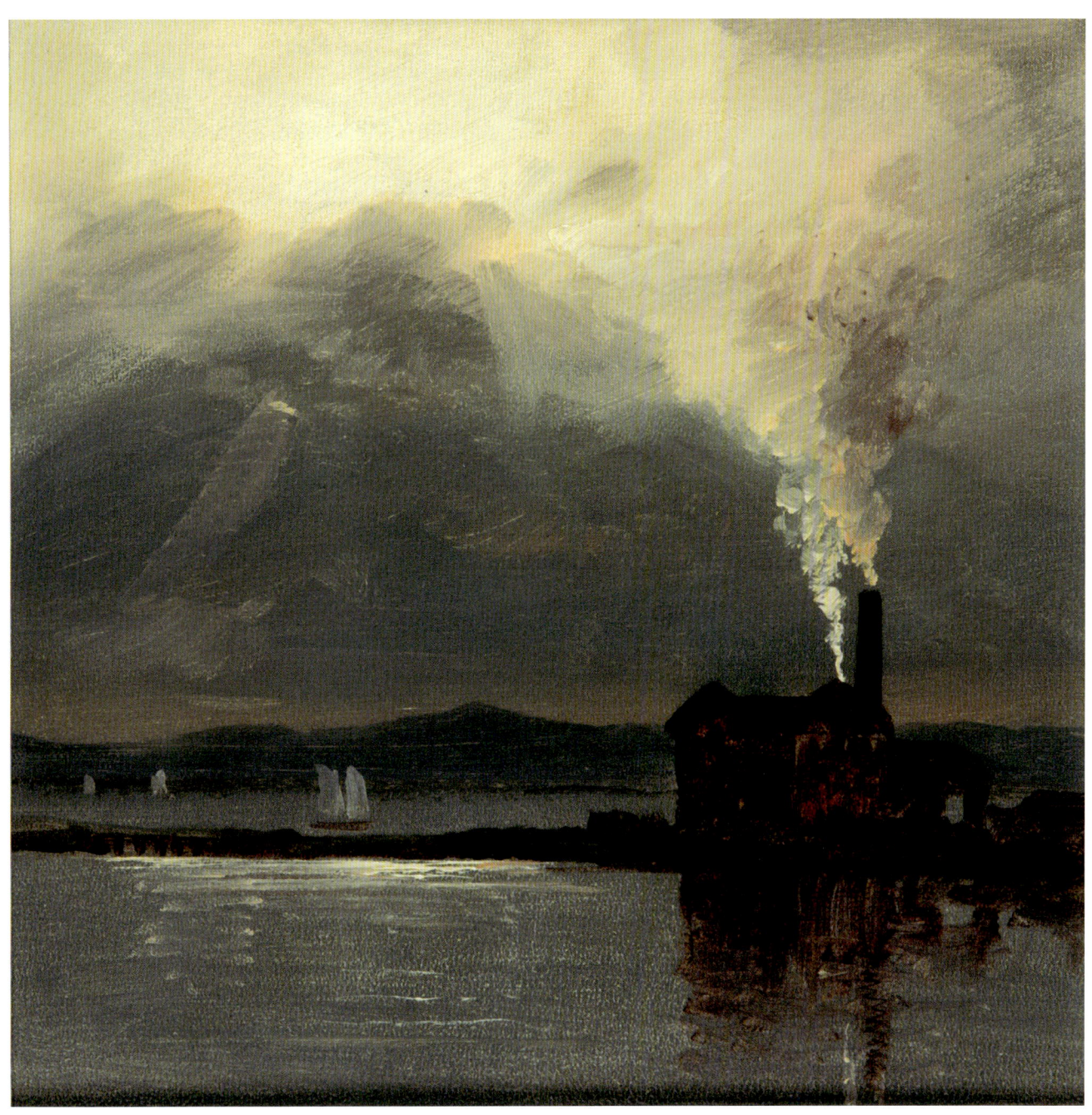

Fig. 9. Frederic Edwin Church (1826–1900). *Hudson River with Factory by Moonlight*, 1844–45. Brush and oil on paperboard, 8 11/16 × 8 15/16 in. (22.1 × 22.7 cm). Cooper Hewitt, Smithsonian Design Museum, Smithsonian Institution, Gift of Louis P. Church (1917-4-44)

Tuckaho marble came from a larger marble vein that was also quarried in upper Manhattan and in Hastings-on-Hudson in Westchester, as documented by Samuel Colman in 1875 (fig. 40). It was used in the construction of important buildings in New York City (such as the Tweed Courthouse) and country (the Washington Monument).[13] After New York's City's Great Fire in 1835, new buildings were required to be built from fireproof materials such as stone and brick, most of which was excavated and processed along the Hudson. These industries thrived not only because of the mineral wealth of the region but also because of the abundance of wood needed for the kilns used to produce bricks and to process limestone for making cement. All of the nineteenth-century paintings in this exhibition feature a shoreline significantly stripped of forest. When Joseph Vollmering painted *View on the Hudson Near Sing Sing, New York*, he chose a distanced vista that characterizes the cluster of neat buildings as a small, self-contained city (fig. 8). The tall chimneys suggest the productivity of the residents, but the scene gives little indication of the strenuous labor and cruel contractors that were part of it. Nevertheless, Vollmering hints at this work by placing broken boulders in the lower right corner. The pastoral tone of the painting may seem odd at first, but carceral institutions of this time were frequently located in natural settings, which were believed to enhance the therapeutic experience of the prisoners. Travel writer Nathaniel P. Willis called Sing Sing "a picturesque object" and included a view like Vollmering's in his illustrated guidebook *American Scenery*.[14]

In addition to marble, iron forged and cast into fences, railings, and architectural elements was shipped downriver to New York City from foundries such as the one depicted in Frederic Edwin Church's *Hudson River with Factory by Moonlight* (fig. 9). Church is one of the best-known members of the Hudson River School, but in this oil sketch, which likely was never meant to be exhibited, he does not aestheticize the environment. The small, soot-encrusted building sits on a promontory reaching out into a still river against a background of modest hills. The smoke belching from the chimney mixes with the clouds in the uncannily illuminated sky, but the result is hardly sublime. This study was likely made as preparatory for a more polished exhibition piece, but its humble nature should not suggest that iron foundries were an undesirable artistic subject. Several of Church's contemporaries produced ambitious landscapes that included them. Johann Hermann Carmiencke's *Poughkeepsie Iron Works (Bech's Furnace)* is one example (fig. 36). Nineteenth-century patrons of art were often New York elites whose wealth relied on upriver factories, which is just one reason such subjects were popular. Carmiencke's view is believed to have been commissioned by Edward Bech, the Danish-born owner of the plant. The painting may have been intended for Bech's local estate, Rosenlund.[15]

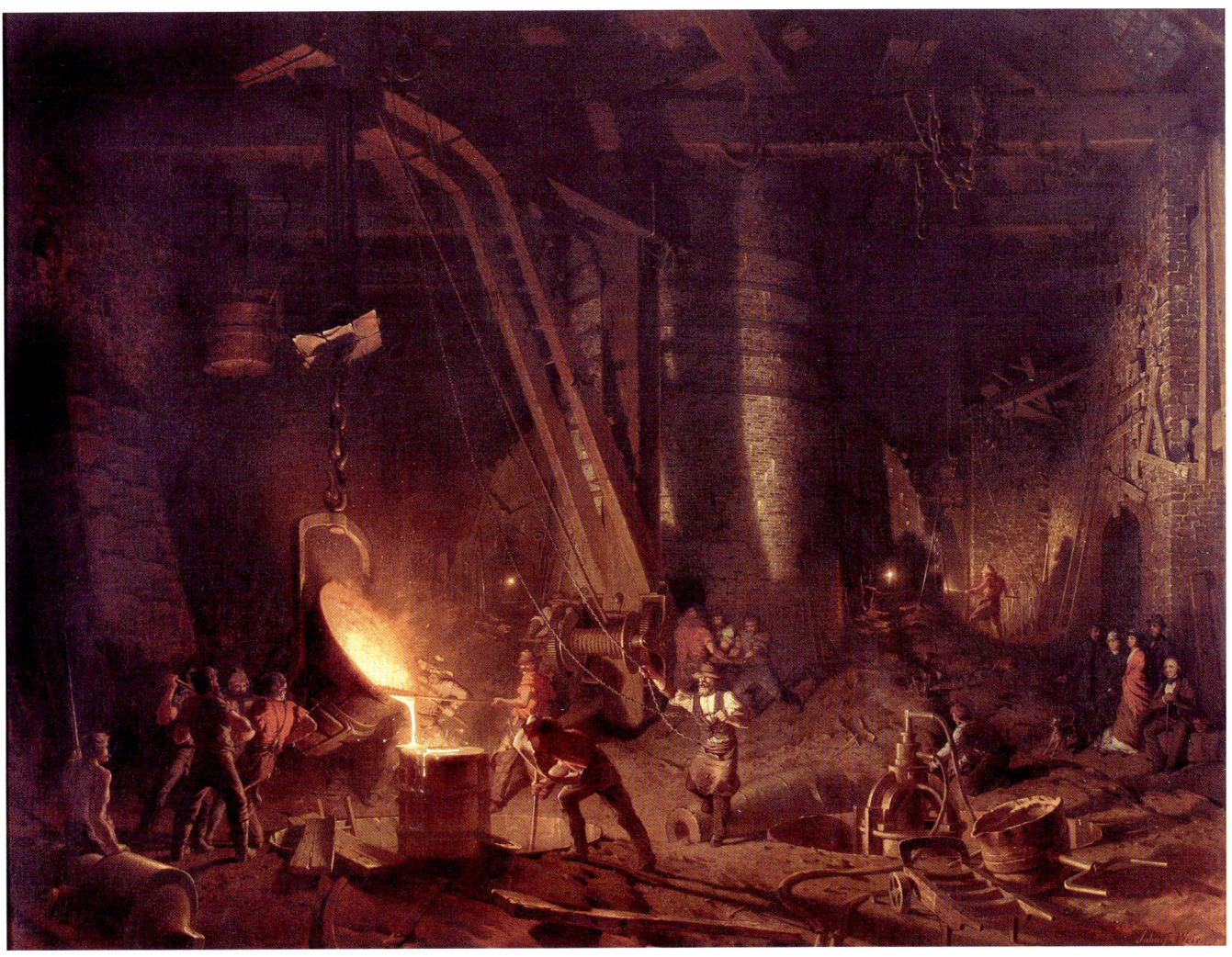

Mining the abundant iron ore deposits of the Hudson River Valley began in the 1730s, peaking during the Civil War. To facilitate transportation of finished iron goods by ship and, later, railroad, many foundries were located along both shores of the river from Westchester to Rensselaer County. Henry Ary's *View of the Hudson* demonstrates how the city grew up around waterfront foundries (fig. 35). The composition brings the eye across the pastoral foreground to the dense industrial area on the far side of the marsh at the center. Although the buildings are diminutive, the smokestacks are prominent, as are the multistory warehouses to their left, both indicating the vitality of this deep-water port. Hudson had begun as Claverack Landing, when brothers from Nantucket moved there to evade British restrictions on American whaling. Local enterprise included wharves, warehouses, and buildings for the processing of spermaceti, and soon the city was a political

Fig. 10. John Ferguson Weir (1841–1926). *The Gun Foundry*, 1866. Oil on canvas, 65 × 80 in. (165.1 × 203.2 cm). Putnam History Museum, Cold Spring, New York

and economic center.[16] As in other towns on the river, Hudson's merchants benefited from chattel slavery, selling both enslaved people and goods produced by them on New York farms and Caribbean plantations. When whaling and seal fishing waned in the first half of the nineteenth century, textile mills, cement factories, and brickworks took their place.

The central building in Ary's view is the Hudson Ironworks, owned by Ary's friend Elihu Gifford, father of painter Sanford Gifford, who had been involved in several industries in the city since moving there in the early 1820s. It is likely the foundry in Church's sketch is also this one; Sanford Gifford and Church were friends and colleagues. One of several foundries in the city, Hudson Iron Works produced as much as 22,000 tons of pig iron a year.[17] Like Carmiencke, Ary nests the industrial buildings neatly into an urban landscape, demonstrating how closely residents lived and worked alongside industry.

John Ferguson Weir's *The Gun Foundry* gives a sense of the conditions inside such buildings (fig. 10). Weir made three paintings at the West Point Foundry, located in Cold Spring opposite its eponymous military installation. Established in 1818, the foundry produced forged and cast-iron products, including guns. Weir's painting demonstrates the sublime aspect of modern industry and reminds us of the public fascination with the sites of production. Like prisons, foundries were a subject of interest to tourists. The West Point Foundry in Cold Spring and the Hudson Iron Works are included in Wade and Croome's *Panorama of the Hudson River from New York to Albany*, a portable map depicting significant buildings along both shores of the river that was frequently carried by steamboat travelers.[18]

As this essay has argued, artworks reveal diverse attitudes toward the Hudson River and record the environmental changes that resulted from the dominance of industry. The Carmiencke and Ary views of iron factory towns build on earlier artists' visions of the region's abundance and productivity. This promise is, however, belied by the inclusion of cleared fields populated with isolated, decorative trees. Industrial production along the Hudson had long relied on charcoal made from local wood. By the time these paintings were executed, there was insufficient timber to fire the forges and coal had to be imported from the west by the Delaware and Hudson Canal Company, which connected Albany to the "rock coal" yards of Pennsylvania.[19] Ironically, clearing the riverbanks of trees had facilitated the erection of the Hudson River Railroad, which itself was built with iron produced using the remnants of the old growth forests. By the mid-nineteenth century, Henry Hudson's dream of harvesting local resources had come true, but artists of the next generation had to find new ways to depict the river's changing industrial and environmental realities.

Notes

1. Memoir of Johannes de Laet (1625), excerpted from J. Franklin Jameson, *Narratives of New Netherland, 1609–1664* (New York: Charles Scribner's Sons, 1909), cited in Roland Van Zandt, ed., *Chronicles of the Hudson: Three Centuries of Travel and Adventure* (Hensonville, N.Y.: Black Dome, 1992), 19.
2. Ibid., 18.
3. Mark Kurlansky, *The Big Oyster: New York on the Half Shell* (New York: Ballantine Books, 2006), 38.
4. Ibid., 62
5. Ibid., 118.
6. Laurel Thatcher Ulrich, *The Age of Homespun: Objects and Stories in the Creation of an American Myth* (New York: Alfred A. Knopf, 2001), 341. On Mohican history, see https://www.mohican.com/brief-history.
7. *The Long Journey Home: The Return of New York's Papscanee Island to the Stockbridge-Munsee Community*, https://storymaps.arcgis.com/stories/4b5d61785b064ff49ceff158e05e89fb.
8. Peter Kalm, *Travels into North America; Containing Its Natural History, and a Circumstantial Account of Its Plantations and Agriculture in General, with the Civil Ecclesiastical and Commercial State of the Country, the Manners of the Inhabitants, and Several Curious and Important Remarks on Various Subjects. The Second Edition. In Two Volumes*, trans John Reinhold Forster (London: T. Lowndes, 1772), vol. 1, 199.
9. Sung Bok Kim, *Landlord and Tenant in Colonial New York: Manorial Society, 1664–1775*. (Chapel Hill: University of North Carolina Press for the Institute of Early American History and Culture, 1978).
10. Gregory Nobles, *John James Audubon: The Nature of the American Woodsman* (Philadelphia: University of Pennsylvania Press, 2017).
11. "State Prison Labor," *Hudson River Chronicle* (Ossining), March 21, 1843.
12. John Sears, *Sacred Places: American Tourist Attractions in the Nineteenth Century* (Amherst: University of Massachusetts Press, 1998), ch. 5.
13. Nathaniel Parker Willis, *American Scenery; or, Land, Lake, and River* (London: George Virtue, 1840), vol. 1, 47.
14. On Tuckahoe marble, see Louis Torres, *Tuckahoe Marble: The Rise and Fall of an Industry in Eastchester, New York, 1822–1930* (Harrison, N.Y.: Harbor Hill Books, 1976).
15. "Bech Family History." Bech Family Homepage, archives.marist.edu/MHP/bechfamily/bechfamilyhomepage.html.
16. Anna R. Bradbury, *History of the City of Hudson* (Hudson, N.Y.: Record Printing and Publishing Co., 1908).
17. Ibid., 179.
18. William Wade and William Croom, *Panorama of the Hudson River from New York to Albany* (New York: J. Disturnell, 1846).
19. https://en.wikipedia.org/wiki/Delaware_and_Hudson_Railway.

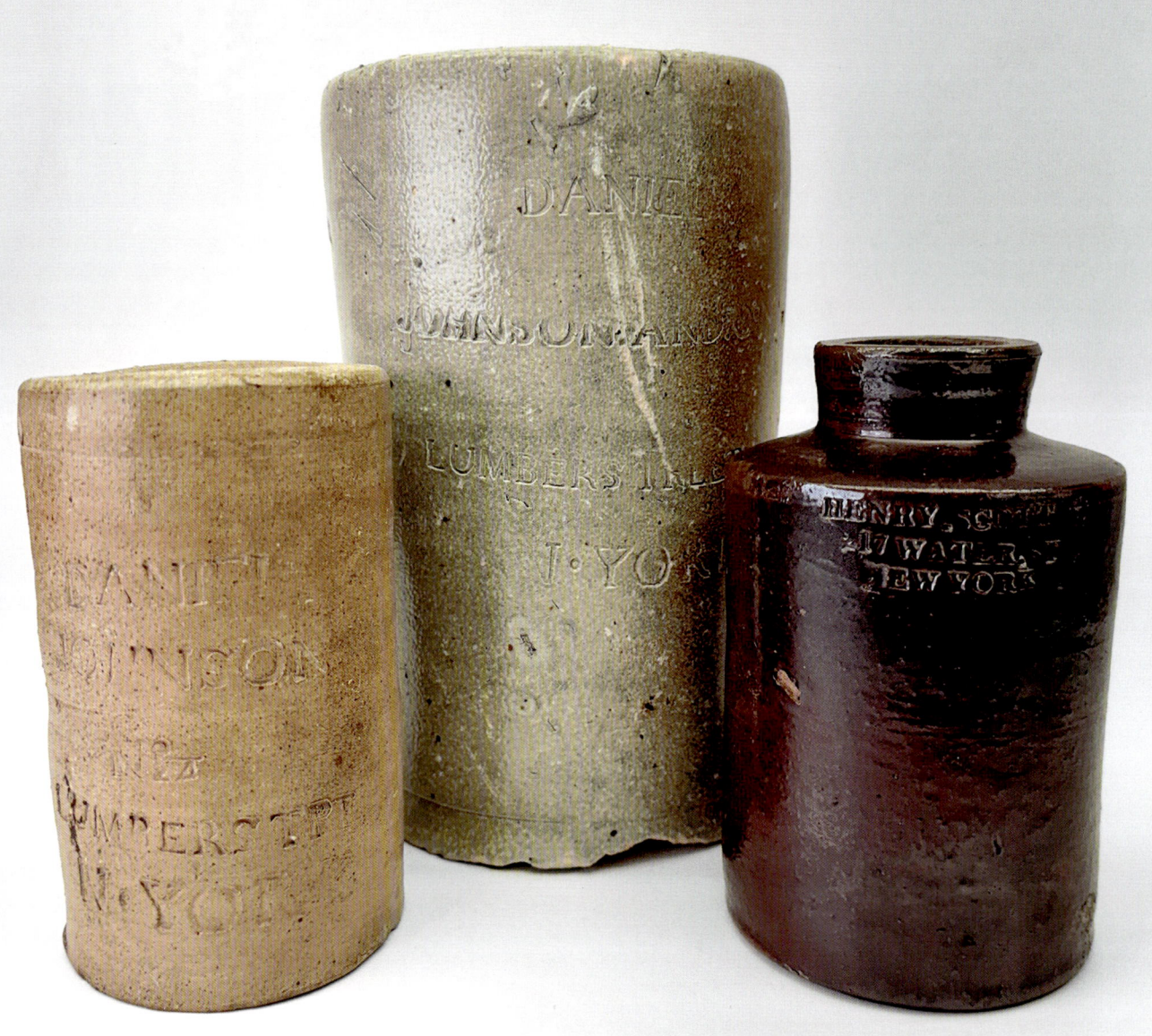

Early New York Oyster Jars

Christopher Pickerell

Among the most distinctive and unique ceramic objects produced in North America during the late eighteenth and early nineteenth centuries are the stoneware oyster jars made for oystermen working in lower Manhattan (fig. 11). These small, unassuming vessels have been recovered by scuba divers and pot hunters from Charleston, South Carolina, and Bermuda to the West Indies and port cities along the coast of South America. These enduring touchstones speak to the rich natural and complex social history of the region and offer an unparalleled glimpse into trade practices and race relations in early Manhattan.

The oysters that filled these jars were harvested from the once pristine and highly productive waters of New York harbor. Here, shellfish flourished for millennia where the nutrient rich brown waters of the Hudson mixed with the briny green waters of the Atlantic over shallow flats ringing the western bank of the river in an area that has long since been filled and now comprises Liberty Park and the Port Jersey Marine Terminal. The first people to live here, the Lenape, managed these grounds for generations before Europeans arrived, harvesting the oysters for their own use and to trade with inland tribes up the Hudson. Ancient shell middens from Manhattan to Dobbs Ferry bear silent witness to this long-standing practice.

The first oyster jars were developed in New York following the English tradition of pickling oysters in small wooden kegs that held a quart, or "hundred count," of oysters. In New York ceramic jars were adopted as a convenient and probably cheaper alternative to the traditional wooden form. In time, the capacity trended smaller as single-serving jars were developed specifically for shipment and trade to the sugar islands of the West Indies and other ports. In 1792 a merchant in Charleston advertised the sale of "pickled oysters in small jars" that had been "just received from New York."[1]

About the same time these small jars were becoming popular, African American watermen rose to prominence in the oyster trade. While the first foreign men of color to pull oysters from local waters did so for the benefit of their Dutch enslavers, by this time free Blacks willingly entered the trade given the economic opportunity it offered. As with other inshore fisheries, a man's earning potential was limited only by the abundance of the target species and the effort expended. In this way, a skilled and determined oysterman

Fig. 11. Left to right: Thomas W. Commeraw (ca. 1772–1823). Oyster jar marked "Daniel Johnson and Co.," 1799–1804. Ceramic with salt glaze, 5¾ × 3½ in. (14.6 × 8.9 cm). Thomas W. Commeraw (ca. 1772–1823). Oyster jar marked "Daniel Johnson and Co.," 1799–1804. Ceramic with salt glaze, 9 × 5¾ in. (22.9 × 14.2 cm). Oyster jar marked "Henry Scott," 1820–40. Ceramic with Albany slip glaze, 6 × 3¾ in. (15.2 × 9.5 cm). Collection of Chris Pickerell

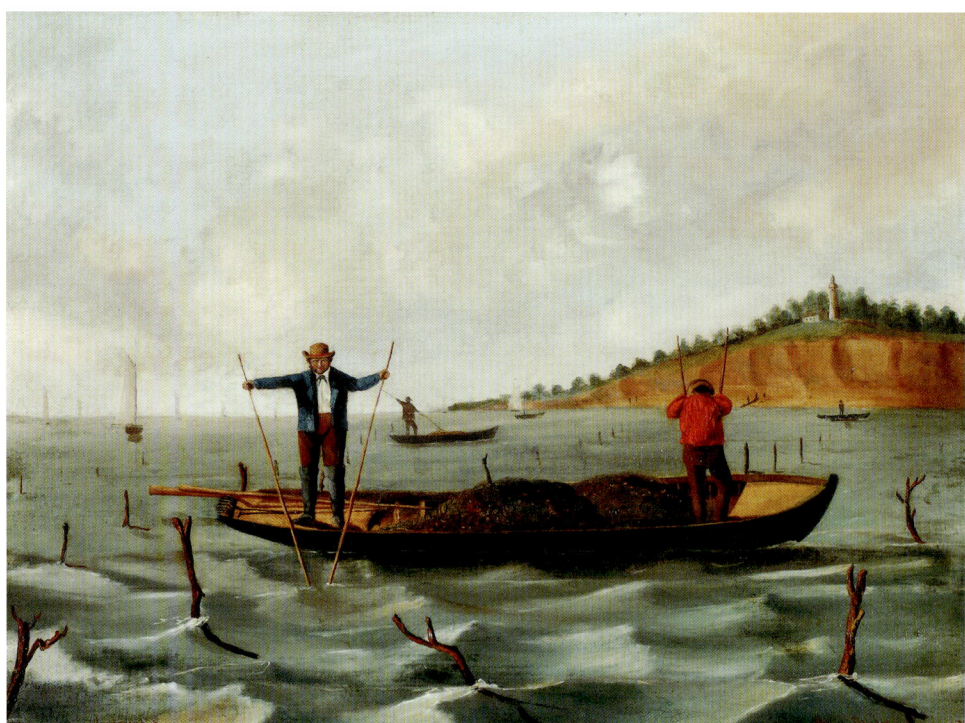

Fig. 12. Alex Matthew (19th century). *Oystering at Prince's Bay*, ca. 1853. Oil on canvas, 25⅛ × 33¼ in. (63.8 × 84.5 cm). Collection of Historic Richmond Town (P01.0057)

could support his family, and if he worked hard enough, might even be able to rent a small stand where he could operate a retail business.

Pioneers in the oyster business included men like Daniel Johnson, George White, and George Brown, who will forever be immortalized by surviving jars imprinted with their name and address made by Black potter Thomas Commeraw working at his shop near Corlears Hook. Commeraw appears to have had a close relationship with these men and produced the only oyster jars with any markings during the first decade of the nineteenth century. It is from this group that we find the largest number of surviving jars in public and private collections. In most cases, the jars were recovered in Guyana and Suriname, South America, areas where the most brutal form of chattel slavery was practiced.

Black oystermen ran their shucking and pickling operations out of cellars under the buildings being constructed along the ever-expanding shoreline. These oyster cellars attracted a diverse and often rough clientele, including prostitutes, gamblers, and petty criminals. Conditions in these cramped and dank basements were often very poor since they were built into recently deposited fill that was dumped into the Hudson and East Rivers. Ironically, it was such filling that would partially lead to the demise of the oyster industry.

Indeed, soon after the turn of the nineteenth century, when Blacks dominated the trade, the wild oyster harvest began to falter. A combination of habitat loss through filling, pollution, and overharvest meant that the traditional oyster grounds could no longer keep up with demand. To overcome these limitations, several white oystermen began to import oyster seed from the Chesapeake and lay it onto recently enclosed and privately controlled oyster grounds off Staten Island and elsewhere. *Oystering at Prince's Bay* (fig. 12) depicts Black oystermen raking within these staked beds, though it is not clear if they were working on their own behalf or, more likely, harvesting for a planter. By this time, the capital-intensive shift in the industry meant that most Blacks were forced to leave the trade, and many transitioned to menial service jobs.

The final chapter for the Black oystermen in New York coincides with the introduction of a new version of the oyster jar, introduced in the 1820s, that included an improved collar to ease closure, better protecting the contents during shipping. These sturdy jars served the last generation of Black oystermen like Thomas Downing of 5 Broad Street and the less well-known Henry Scott of 117 Water Street, who left their mark on the oyster trade with surviving jars of their own. These jars are the only traces of the men who made their living trading in one of New York's most important foods.

Note
1. Advertisement for John G. Mayer in *City Gazette*, December 5, 1792.

2.ᵉ Livraison • AMÉRIQUE SEPTENTRIONALE • ÉTAT DE NEW-YORK. • Pl. 2

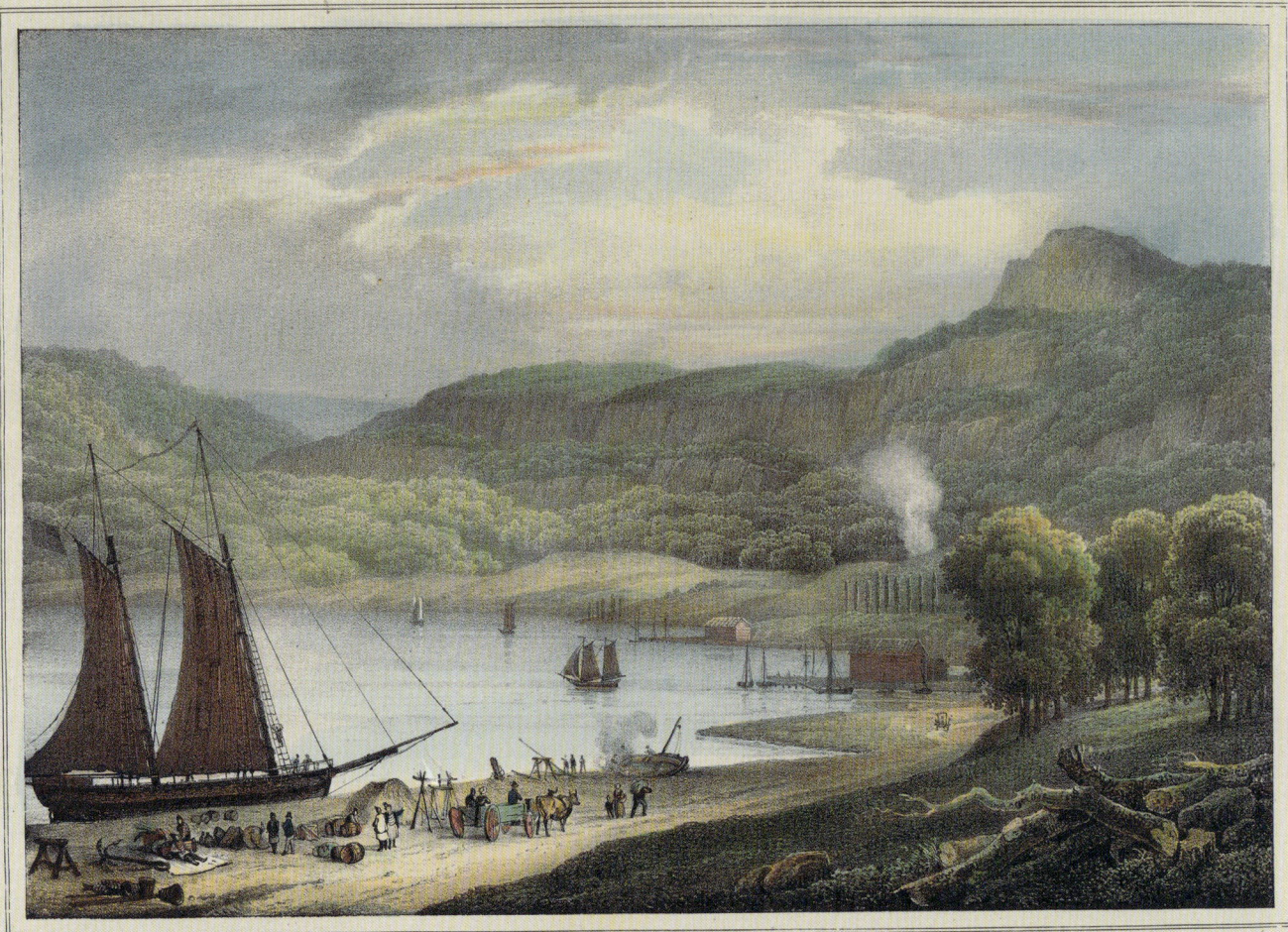

Lithographié par Bichebois fig. par V. Adam.   Dessiné d'après nature par J. Milbert.

Port d'Haverstraw ou de Warren.   N.º 6   Haverstraw or Warren landing.

Haverstraw sive Warren portus.   Der hafen Haverstraw oder Warren.

Paris, chez E. Ardit, éditeur rue Vivienne N.º 2

Landscape Painting and the Hudson Valley Brick Industry

Ross Barrett

As they painted the developing Hudson River Valley, nineteenth-century artists consistently relegated one prominent local industry to the margins (or blindspots) of their compositions: brickmaking. Brick manufacturing had deep roots in the Hudson Valley.[1] Colonial settlers first encountered the region's strong blue clay deposits—left on the edges of the Hudson's riverbed by the retreat of Pleistocene Era glaciers and used by local Lenape communities to make pottery, utensils, and other implements since the Middle Woodland Period (ca. 200 BCE to 300 CE)—in the 1630s.[2] By the Revolutionary period, artisans had erected crude brickyards throughout the valley. As a lithograph of James Wood's early brickyard at Haverstraw suggests (fig. 13), brickmaking remained a small-scale endeavor into the early nineteenth century. Taken together, the small sloop at left, isolated plume of kiln smoke at right, and distant heavily wooded slopes attest to the modest productivity and still-rustic character of the region's brick operations.[3]

The Hudson Valley brick industry expanded dramatically in the 1840s and 1850s. New York City's rapid growth and adoption of new construction codes after the Great Fire of 1835 created a voracious market for fireproof building materials.[4] Rushing to meet this demand, local entrepreneurs modernized the valley's brickyards, introducing new processes (such as the tempering of clay with anthracite coal) and machines (such as Richard Ver Valen's automatic molder) that streamlined the brickmaking process and boosted the productivity of area firms.[5] These innovations transformed the region's artisanal brickmaking business into an explosively productive industrial sector: one scholar estimates that Hudson Valley yards employed at least seven thousand workers and produced one billion bricks per year by the end of the century.[6]

The artists who wandered the Hudson Valley in the middle decades of the nineteenth century would have been surrounded by the sights, sounds, and smells of industrial brickmaking. Few chose to include any signs of the business in their pictures of the region. Consider the artful evasions that structure period views of Haverstraw, a village on the western bank of the Hudson that was both a popular artistic destination and a brickmaking hub. In the 1860s and 1870s John Frederick Kensett, Francis Silva, John William Hill,

Fig. 13. After Jacques Gerard Milbert (1766–1840). *Haverstraw or Warren Landing*, 1828–29. Lithograph, 7¹¹⁄₁₆ × 11³⁄₁₆ in. (19.5 × 28.4 cm). Yale University Art Gallery, New Haven, Mabel Brady Garvan Collection (1946.9.1947)

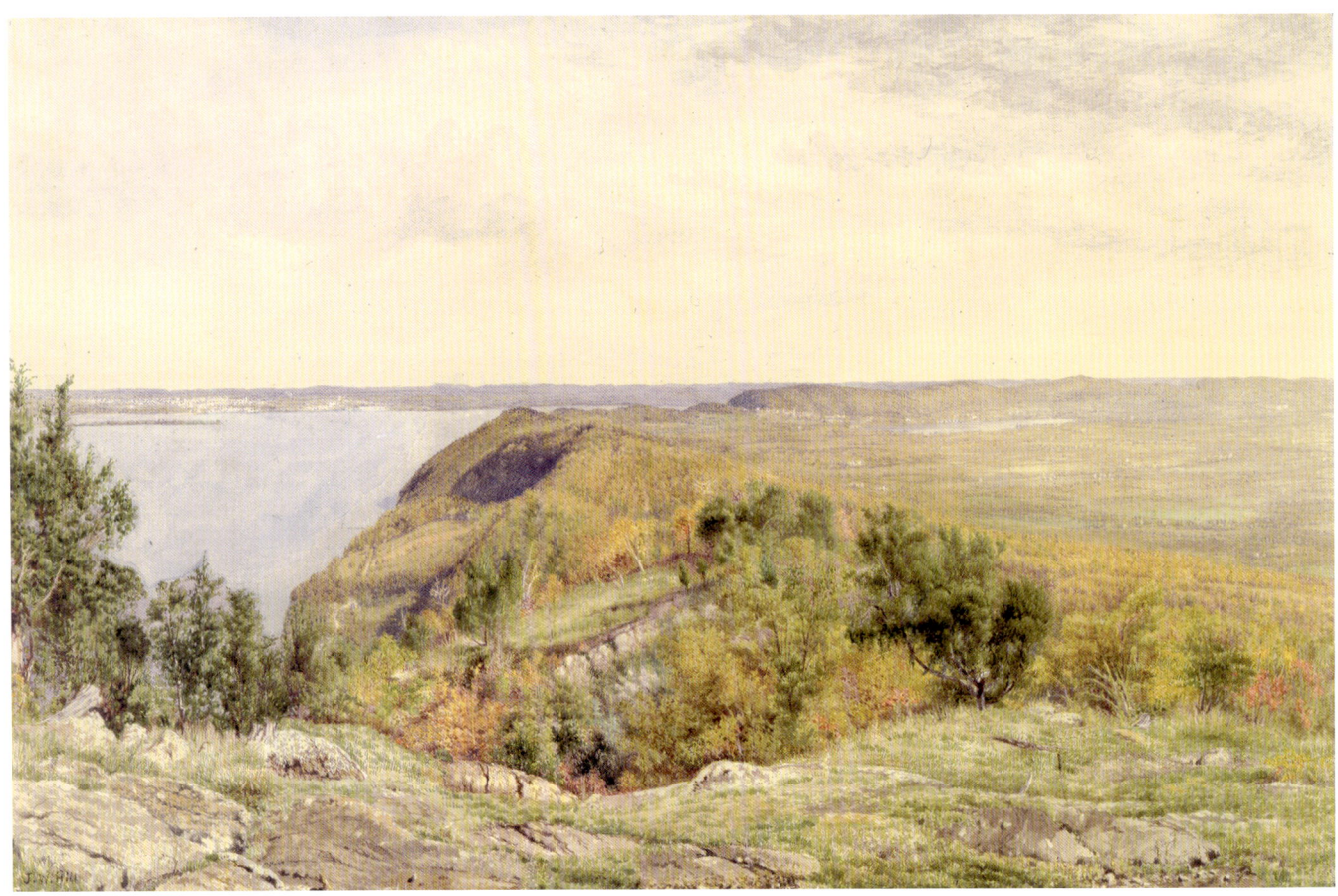

Fig. 14. John William Hill (1812–1879). *View from High Tor, Haverstraw, New York*, ca. 1866. Watercolor and graphite on paper mounted to card, 11¾ × 18 in. (29.9 × 45.7 cm). New-York Historical Society, Purchase, Foster-Jarvis Fund

Sanford Gifford, and others composed major canvases in and around Haverstraw; their works almost unfailingly elided the dozens of brickyards that spread out along the village's waterfront. Some focus on the river: Gifford's *Haverstraw Bay* (1868) and Silva's *On the Hudson Near Haverstraw* (1872), for example, show scenes of sailing and fishing that accentuate the Hudson's breadth and minimize its industrial shorelines. Others feature extensive views that conceal Haverstraw's brickyards behind intervening landforms; Hill's *View from High Tor, Haverstraw, New York* (fig. 14) thus sets a grassy ridge between the viewer and the village's brickmaking fringe.[7]

The brick industry's absence from these and other period landscapes was no accident. As a dangerous business that is all, in the words of Herman Melville, "mud and mire," brickmaking would have struck most middle-class viewers as an inappropriate subject for painterly treatment.[8] Haverstraw's brick business carried a particularly alarming set of social and environmental associations for this audience. The industrialization and de-skilling of brickmaking drew an influx of Irish immigrants to the village in the 1840s and 1850s; these workers would be joined at the turn of the nineteenth and twentieth centuries by Italian immigrants and African American migrants from the South.[9] Haverstraw's immigrant and migrant workers fashioned cultures of recreation and rebellion that the local native-born gentry found deeply troubling. Mid-nineteenth-century newspapers from the region are filled with hysterical accounts of brickmakers' real or imagined "drunkenness," "intemperate habits," and "disorderly conduct."[10] Brick workers' willingness to challenge authority and protest poor working conditions were even more alarming to middle-class observers; when Haverstraw brickmakers struck in May 1853 for a ten-hour workday, for example, a *Rockland County Journal News* editorial decried the protest as an "unlawful and inflammatory . . . row."[11]

Others worried about the toll that Haverstraw brickmaking was taking on the natural environment. Even the most fervent boosters acknowledged that the industry's rapid expansion had irreparably altered Haverstraw's landscape. Noting that the town's "gentle slope and green turf have been removed by the brickmaker's spade," one 1861 editorial admitted that Haverstraw had taken on "a broken and gray appearance from the river."[12] Local residents realized as early as the 1840s, moreover, that emissions from Haverstraw's brickyards were spoiling local farmlands; an 1846 *Farmer's Cabinet, and American Herd-Book* article noted that "gases from burning brick kilns" had a "disastrous effect" on area orchards.[13]

Although brickmaking's worrying social and environmental associations convinced most nineteenth-century painters to erase Haverstraw's yards from their finished landscapes, the industry does make an occasional appearance in studies and experimental works from the period. Marie-François-Régis Gignoux's *Haverstraw on the Hudson* (fig. 18) speaks to the complex symbolic dynamics that could shape these productions. After beginning his career in France, Gignoux relocated to the United States in 1840 and established himself as a painter of Hudson River views and sublime landscape subjects such as Niagara Falls in the 1850s and 1860s.[14] While celebrated for the "dreamy" and "delightful" effects of his landscape works, Gignoux nurtured an interest in grittily unpoetic subjects.[15] In the late 1850s he made several pictures of New York's Fulton Market, a rundown commercial hub where butchers, fishmongers, sporting men, and sex workers intermingled. Considering these works, one approving critic noted that Gignoux painted the disagreeable market in a way that allowed "its splendors . . . [to] be enjoyed without experiencing the discomforts of its odors."[16]

It was likely this interest in workaday subjects that led Gignoux to undertake *Haverstraw on the Hudson*, which seems to have been conceived as an experimental sketch.[17] Adopting a vantage just outside Haverstraw's downtown, the painting traces a sweeping view that stretches from the village's southern fringe (right) to Pullen's Point (center); Haverstraw's steamboat landing is slightly to the left, flanked by a white paddle wheeler. The expansive composition also includes a small vignette of brickmaking: a brickyard complex with kilns, a smokestack-equipped pug mill, and a warehouse appear on a raw red promontory in the lower right, bracketed by three figures, possibly workers, and a suggestively denuded tree.

Considered in isolation, this vignette might be read as an unusually forthright acknowledgment of brickmaking's presence in Haverstraw and a pointed evocation of the industry's disquieting material realities. But *Haverstraw on the Hudson* implicates the brickyard in a series of oppositions that subtly complicate its meanings. A vignette of sightseeing is directly across the canvas from the brick operation; at lower left, two fashionably attired women—one seated, one standing—gaze out at Haverstraw from a rocky outcropping. A diagonal swath of green farmland—complete with rail fences, shade trees, and grazing cattle—stretches out below the tourists and angles into the distance. Incorporating signs of industry and agriculture, proletarian labor and middle-class property, healthful nature and environmental damage in a single landscape enjoyed by two bourgeois spectators, *Haverstraw on the Hudson* would have been read in its moment as an experiment in picturesque representation—an established landscape mode that harnessed contrasting landforms and evocations of social difference to achieve mixed compositions defined by visual and thematic variation.[18] Viewers familiar with this mode would have understood the brickyard vignette less as a venture in social realism than as an interesting note of visual and social contrast within a stimulatingly multifarious scene. Employing the conventions of the picturesque to aestheticize local brickmaking, *Haverstraw on the Hudson* worked to soften the harsh material realities of industrialization and defuse the troubling implications of Haverstraw's primary business. In so doing, the painting explored an approach to the representation of brickmaking that other period artists and writers would sometimes adopt.[19]

Notes
1. On Hudson Valley brickmaking, see Richard O'Connor, "A History of Brickmaking in the Hudson Valley," Ph.D. diss., University of Pennsylvania, 1987; George Hutton, "The Zenith and Sudden Decline of the Great Hudson River Brick Industry," *Hudson Valley Regional Review* 19 (2002): 17–29; and George Hutton, *The Great Hudson River Brick Industry: Commemorating Three and a Half Centuries of Brickmaking* (New York: Purple Mountain Press, 2003).
2. Timothy Scarlett, Daniel Rahn, and Daniel Scott, "Bricks and an Evolving Industrial Landscape: The West Point Foundry and New York's Hudson River Valley," *Northeast Historical Archaeology* 35 (January 2006): 31–32; and Lucianne Lavin, *Connecticut's Indigenous Peoples: What Archaeology, History, and Oral Traditions Teach Us About Their Communities and Cultures* (New Haven: Yale University Press, 2013), 150.

3. On Milbert's lithograph, see Philip Weimerkirsch, "Two Great Illustrated Books About the Hudson River: William Guy Wall's *Hudson River Port Folio* and Jacques Gérard Milbert's *Itinéraire pittoresque du fleuve Hudson*," in *Adirondack Prints and Printmakers: The Call of the Wild*, ed. Caroline Martin Welsh (Syracuse: Syracuse University Press, 1998), 26–41; and Laura L. Vookles, "The Whole Scenery on the Hudson River: Journeys and Destinations," in *The Panoramic River: The Hudson and the Thames* (New York: Hudson River Museum, 2013), 98–99.
4. O'Connor, "History of Brickmaking in the Hudson Valley," 8–17.
5. Ibid., 125–40.
6. Ibid., 149–50.
7. Robert Havell's undated *View of the Hudson River at Haverstraw Bay* (Private collection) shapes a similarly distanced view of the village.
8. Herman Melville, "Israel Potter," *Putnam's Monthly* (February 1855): 181.
9. On Italian and African American brick workers, see O'Connor, "History of Brickmaking in the Hudson Valley," 242–43.
10. "Riot and Case of Shooting at Haverstraw," *Rockland County Journal News*, June 3, 1854; "Murder at Haverstraw," *Rockland County Journal News*, June 12, 1858; untitled article, *Rockland County Journal News*, August 27, 1859; and "Increase of Crime in Our County," *Rockland County Journal News*, June 23, 1869.
11. "Great Row in Haverstraw," *Rockland County Journal News*, May 7, 1853.
12. "History of Haverstraw," *Rockland County Messenger*, August 8, 1861.
13. "Deleterious Effects of Brick Yards," in Josiah Tatum, *Farmers' Cabinet, and American Herd-Book* 40 (December 15, 1846): 151.
14. On Gignoux's understudied career, see John K. Howat, *The Hudson River and Its Painters* (University Park: Pennsylvania State University Press, 1983), 155–57; Robert McGrath, *Gods in Granite: The Art of the White Mountains of New Hampshire* (Syracuse: Syracuse University Press, 2001), 86; David Dearinger, *Paintings and Sculpture in the Collection of the National Academy* (New York: Hudson Hills Press, 2004), 226; and Ann Lee Morgan, *The Oxford Dictionary of American Art and Artists* (New York: Oxford University Press, 2007), 180.
15. "Fine Arts," *New York Herald* (February 4, 1863): 4; and "Arts and Artists," *New Mirror* (May 6, 1843): 76.
16. "Art Items," *New York Daily Tribune*, July 14, 1860. On Fulton Market, see Mark Kurlansky, *The Big Oyster: History on the Half Shell* (New York: Ballantine Books, 2006), 188–92.
17. *Haverstraw on the Hudson* is much less finished than the large-scale exhibition works—such as *New Hampshire* (1864; Hood Museum of Art)—that Gignoux executed during the period.
18. On the picturesque, see Ann Bermingham, *Landscape and Ideology: The English Rustic Tradition, 1740–1860* (Berkeley: University of California Press, 1986), 57–87; and John Evelev, *Picturesque Literature and the Transformation of the American Landscape, 1835–1874* (New York: Oxford University Press, 2021), 1–24.
19. Hudson Valley writer Nathaniel Parker Willis, for example, included a pejorative and racialized account of the supposedly picturesque qualities of the region's Irish brickmakers in his 1855 collection *Out of Doors at Idlewild*. After noting that "these miles of brickyards" are a "loose and lively Irish fringe to our quiet American neighborhood," Willis averred that "the Irish [brickworkers] are a variety that comes in well for contrast and invigoration to the musing and half-conscious picture formed upon the eye during a drive." See Nathaniel Parker Willis, *Out of Doors at Idlewild; or, the Shaping of a Home on the Banks of the Hudson* (New York: Scribner's 1855), 397–98.

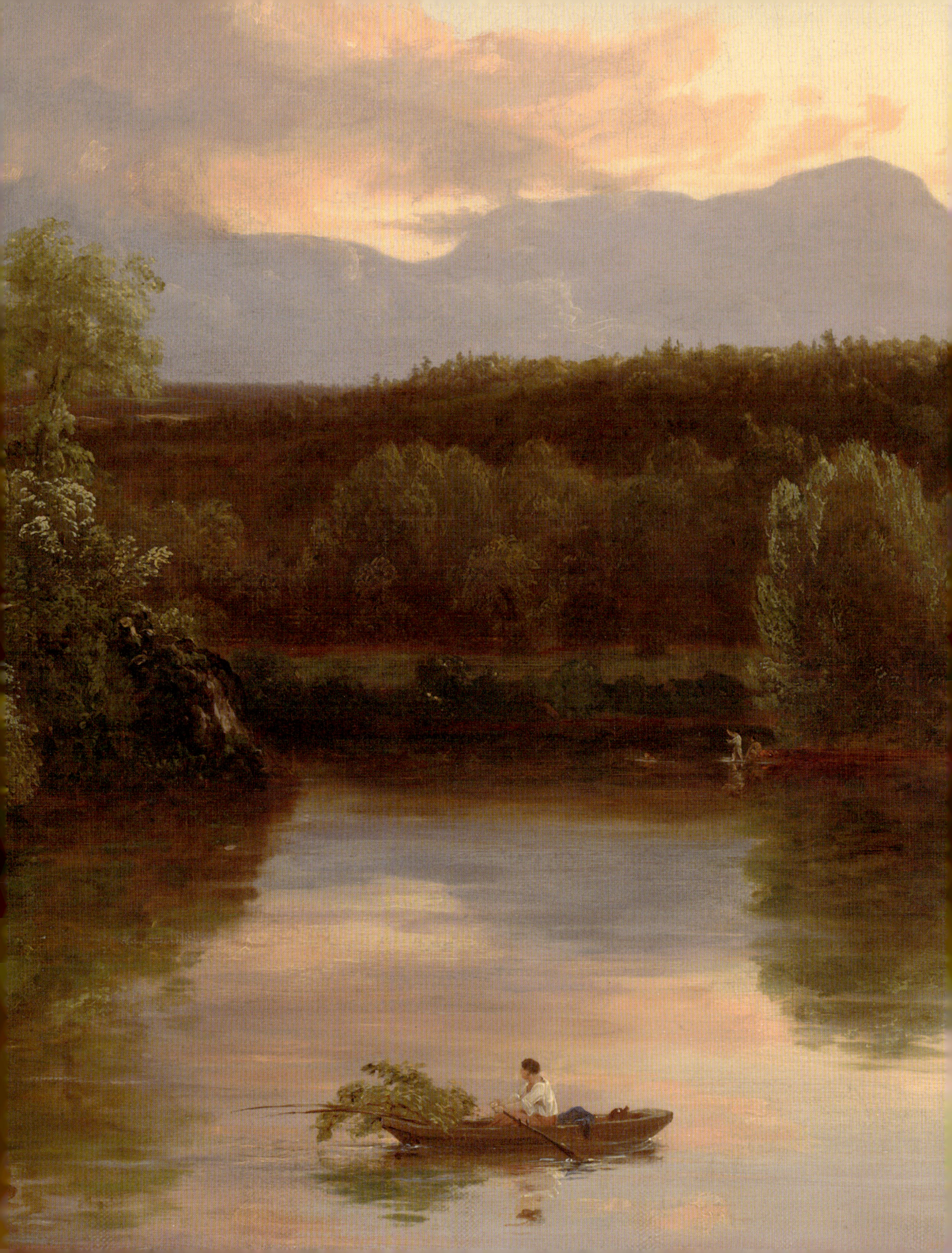

Omissions from the Canon

Artists' Representations of Industry along the Hudson River in the 19th and 20th Centuries

Annette Blaugrund

> The river scenery of the United States is a rich and boundless theme. The Hudson for natural magnificence is unsurpassed.
>
> —Thomas Cole, "Essay on American Scenery," 1836[1]

When Thomas Cole, the father of the Hudson River School of landscape painting, visited the Hudson Valley for the first time in 1825, he envisioned a virgin wilderness. By the mid-nineteenth century the river's shores were filled with smoke-spewing factories and railroads that he and other artists excluded, concealed, or telescoped in the distance. He depicted in *North Mountain and Catskill Creek* (fig. 15), among other paintings of this major Hudson River tributary, a carefully observed bucolic shoreline view with several small-scale figures. The man rowing the boat in the foreground carries willow branches used to make baskets; the man in the center tarries as his horse drinks; and another man in the deep background is fishing with a Native American sitting by his side. Apart from these noninvasive occupations, there are no signs of industry or encroaching civilization.[2]

The industries along the Hudson River from 1860 to 1940, the period covered in this essay, included fishing, farming, ice harvesting, ferry and steamboat transportation, leisure activities, brickmaking, mining, iron foundries, sugar refineries, and shipbuilding, among others. Located along the increasingly industrialized shoreline were railroads, ports, apartment complexes, power plants, and prisons such as Sing Sing, in operation since 1826.

At the height of the Hudson River School, ideas about the function of art and concern for the environment resulted in few paintings purposely documenting industry and shoreline erosion. While many images of the river capture a natural paradise, several little-known works show industry along the shore from Albany southward to where the Hudson merges with the Atlantic Ocean, the scope of the *Shifting Shorelines* exhibition.

Thomas Cole. *North Mountain and Catskill Creek* (detail), p. 42

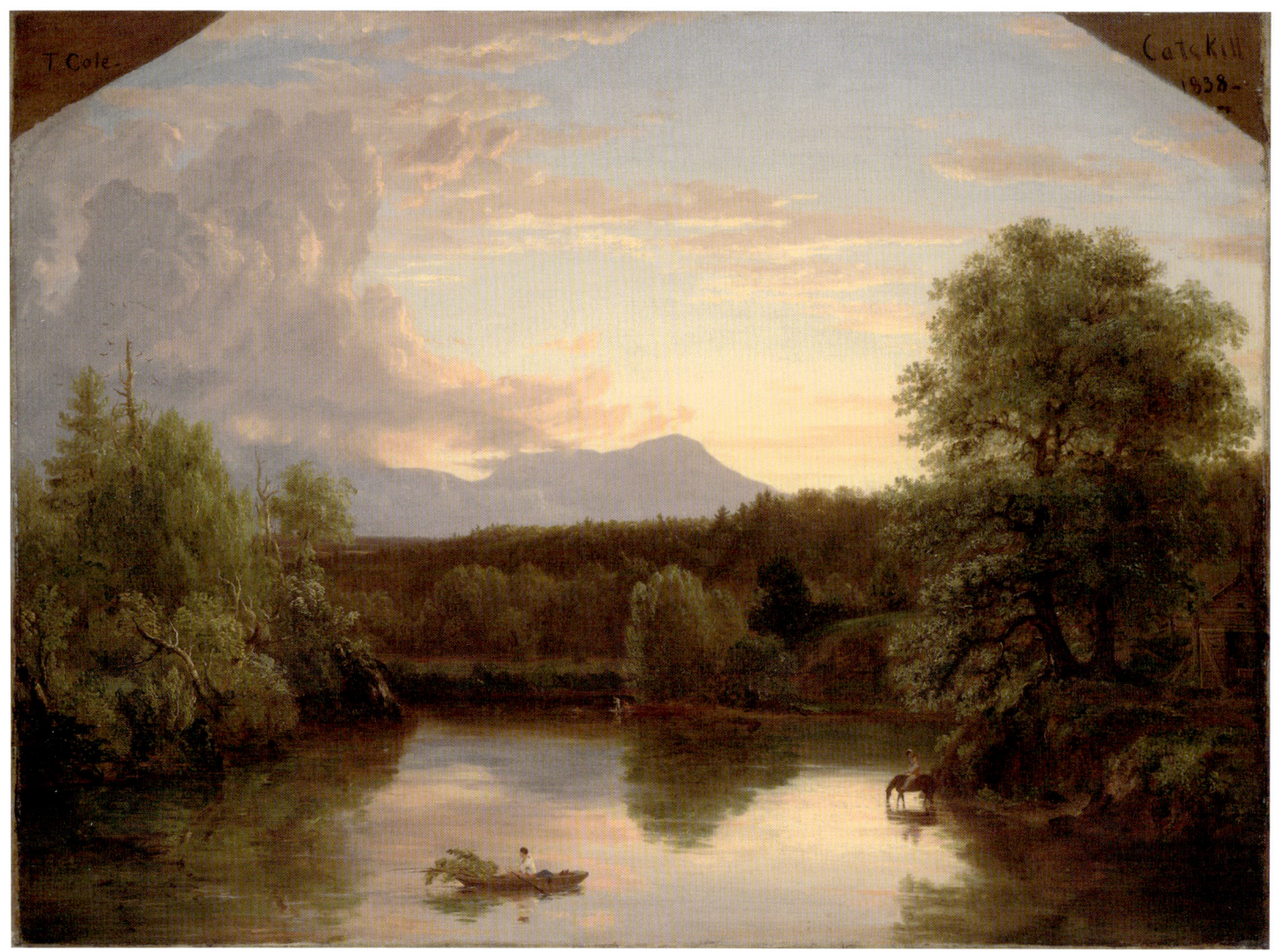

Fig. 15. Thomas Cole (1801–1848). *North Mountain and Catskill Creek*, 1838. Oil on canvas, 26 7/16 × 36 7/16 in. (67.2 × 92.6 cm). Yale University Art Gallery, New Haven, Gift of Anne Osborn Prentice (1981.56)

Which artists painted these subjects or eliminated and obscured them? Was it the artist's choice or that of patrons who wanted sublime wilderness scenes to decorate their walls? Was patronage the reason so many artists avoided industrial images? And what fostered the change at the end of the nineteenth century to inclusion and pride in modern achievements? The answers are varied, reflecting changing attitudes about the Industrial Revolution, artistic evolution, and cultural vicissitudes, as documented in the following examples.

Worthington Whittredge's *Shad Fishing on the Hudson* (fig. 16) illustrates the marshes and fishing, offering a view of shoreline ecology and leisure activity and revealing wetlands that are fundamental to the diverse ecosystems in the Hudson tidal estuary. Saltwater flows daily from the ocean and mixes with freshwater flowing down from northern tributaries. The Hudson estuary was among the most productive ecosystems, teaming with nurseries for fish such as striped bass, sturgeon, and shad that were a plentiful, long-standing food source. Shad's remarkable ability to adapt from saltwater to freshwater made it a significant fish historically, a delicacy popular with Native Americans and later with colonists and people across the country.[3]

The farms along the river also supplied food for the growing metropolis. Julie Hart Beers's *Cows in a Landscape* (fig. 17) shows cows that supplied milk and meat and a cultivated field in a bucolic setting. Beers was one of many women landscape painters working in the style of the Hudson River School. She, Susie Barstow, Eliza Greatorex, Fidelia Bridges, and others exhibited and sold their paintings side by side with their male counterparts at such popular venues as the National Academy of Design but were omitted from art books written exclusively by men.[4]

Several artists focused on the beauty of the river around West Point as did John Ferguson Weir in *View of the Highlands from West Point* (cat. 76), looking north to the mountains. John Weir's father, Robert Walter Weir, was the drawing instructor at the West Point Military Academy. Growing up in the area, Weir became fascinated with the West Point Foundry, established in Cold Spring in 1818, across the river from the school. Situated on the banks of the Hudson, it was the largest and most modern foundry in the United States and the place where most of the cannons for the Union were manufactured during the Civil War. After the war, Weir painted *The Gun Foundry* (fig. 10), revealing the interior's noise, smoke, heat, and grime. As iron was abundant in the area, the beauty of the highlands was impacted by the denuding of the rich forests for lumber needed to fuel the furnaces that produced pipes, railroad engines, cannon balls, and the casting of the Parrott rifle seen in this painting.

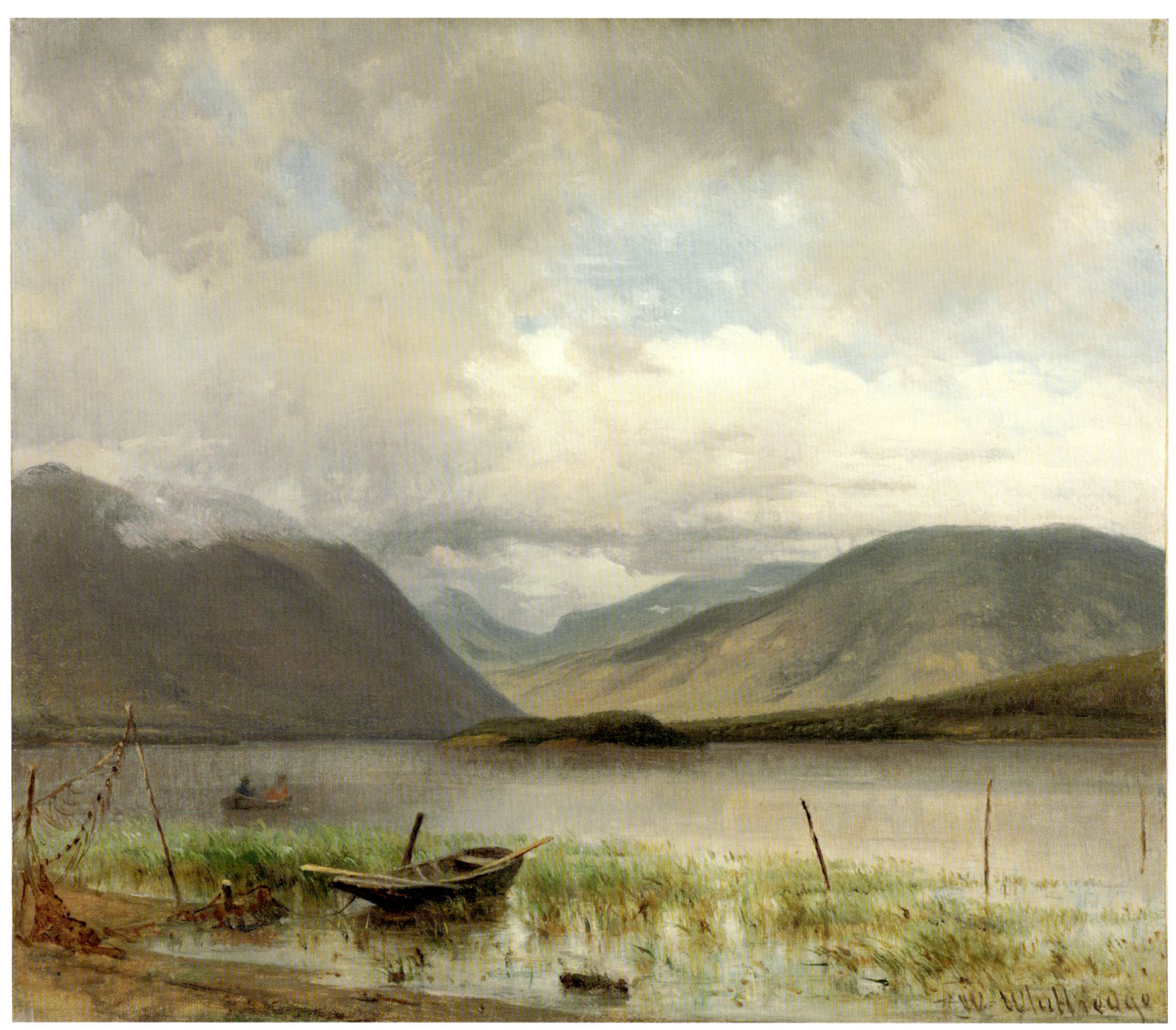

Fig. 16. Worthington Whittredge (1820–1910). *Shad Fishing on the Hudson*, ca. 1875. Oil on canvas, 11 1/2 × 13 1/2 in. (29.2 × 34.3 cm). Philadelphia Museum of Art, Gift of Marguerite and Gerry Lenfest, 2008 (2008-124-3)

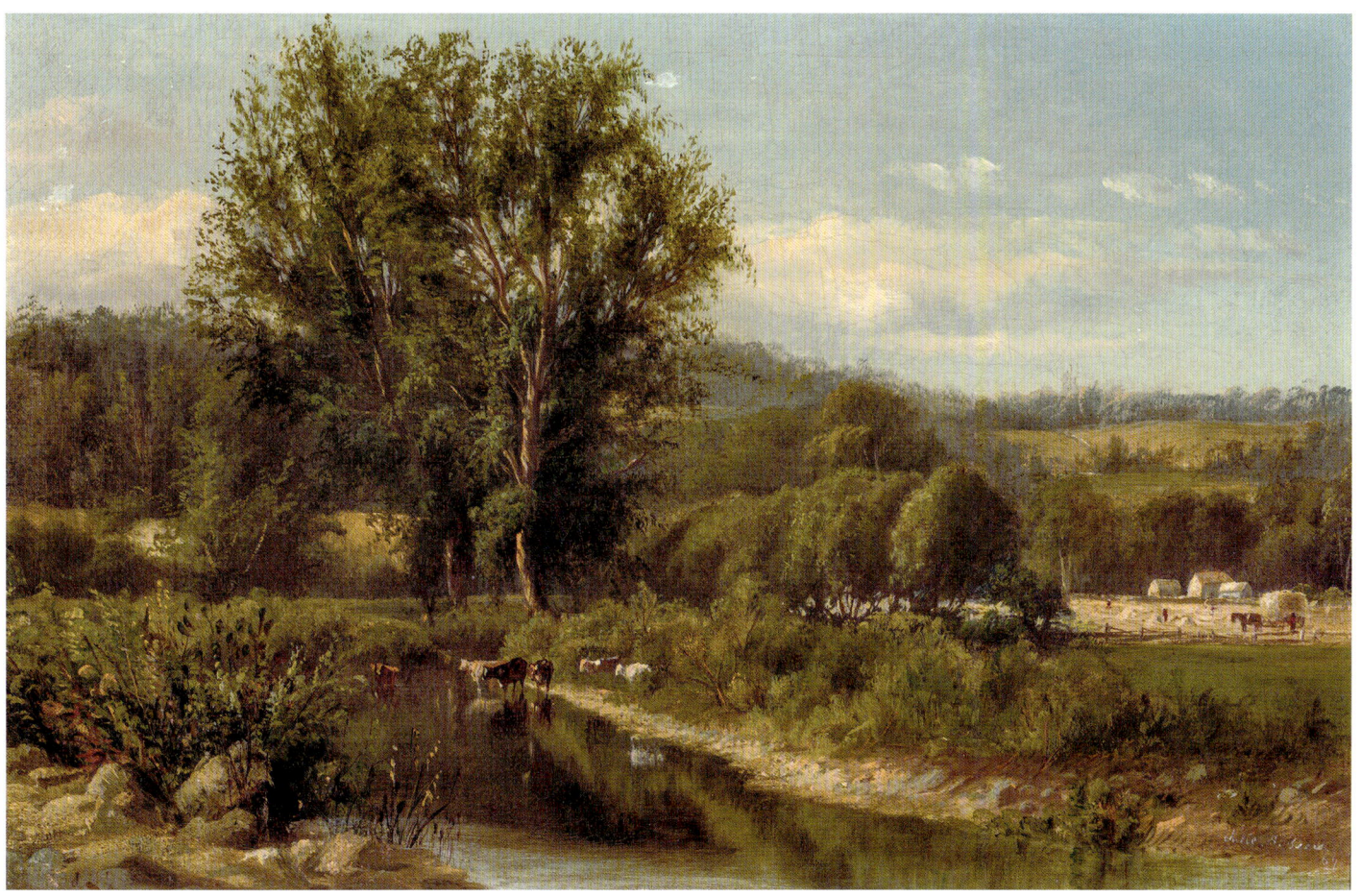

Fig. 17. Julie Hart Beers (1835–1913). *Cows in a Landscape*, 1869. Oil on canvas, 9 1/8 × 14 9/16 in. (23.2 × 37.0 cm). Lebanon Valley College Fine Art Collection, Annville, Pennsylvania (2020.1.45)

In David Johnson's panoramic *View from Garrison, West Point, New York* (cat. 42), a hardly discernable academy is found in the left middle ground. In the nineteenth century, West Point developed a rigorous program of education and military training and became an important cultural and social center. It attracted an elite corps of cadets and influential visitors to its campus. Even less visible is the foundry across the river, reddish in color with a faint plume of smoke emanating from a chimney, perpetuating the vision of an idyllic wilderness that obscured signs of human activity.

A little more than twelve miles south of Garrison is the town of Haverstraw, in Rockland County, New York, located at the Hudson River's widest point. It was often referred to as "Bricktown" because of its brickmaking industry, based on the clay made from the water and the rich soil that lined the town's waterfront. Marie-François-Régis Gignoux's *Haverstraw on the Hudson* (fig. 18) depicts the shoreline where, by the early twentieth century, there were more than forty brickmaking factories. Many of the brick buildings constructed in New York City through the early twentieth century were

OMISSIONS FROM THE CANON

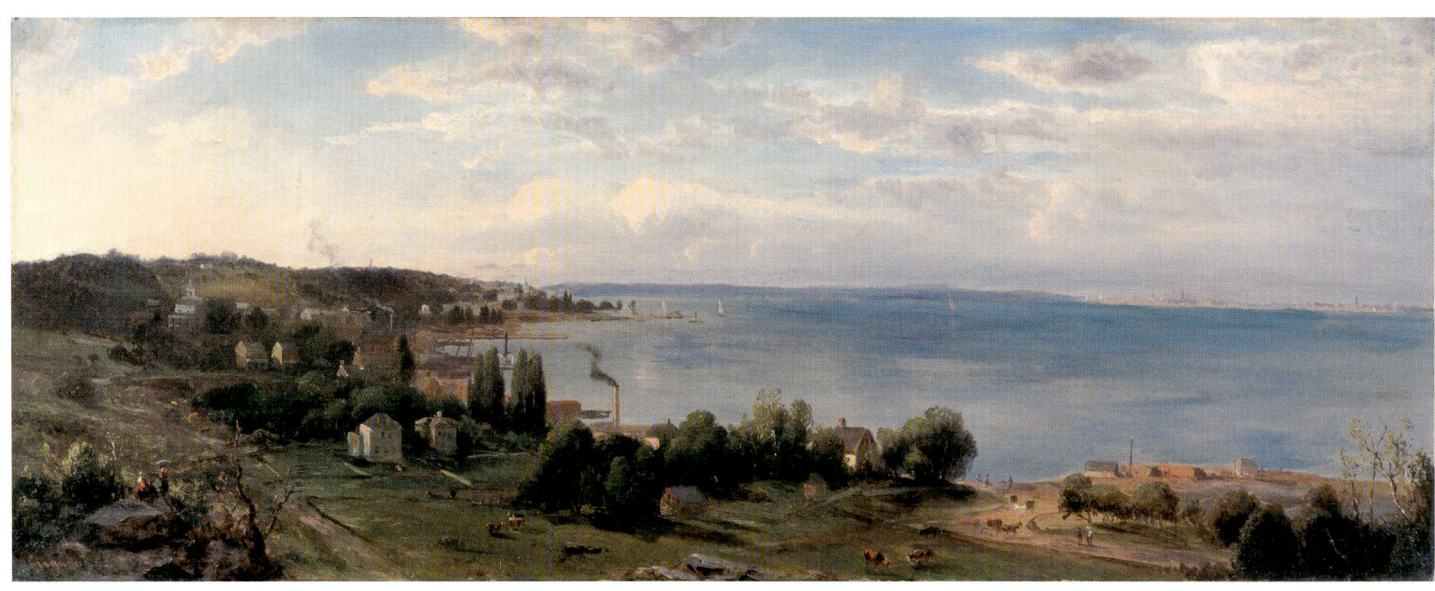

Fig. 18. Marie-François-Régis Gignoux (1814–1882). *Haverstraw on the Hudson,* 1860–65. Oil on canvas, 13½ × 35¼ in. (33.0 × 88.9 cm). Albany Institute of History & Art, Purchase (1951.68)

composed of bricks manufactured in Haverstraw. Gignoux, however, focused on the river rather than on this major polluting industry and its laborers.

Mining was another industry that marred the land bordering the river. In his watercolor *Hook Mountain, Nyack* (fig. 37), Edward Hopper captures the defacement of the mountain by a quarry that had been in operation since 1785. It provided red sandstone that was shipped down the river. The dirty, dusty, noisy mine was bought by preservationists who closed it and created a recreation area in 1911. Hopper's family home in Nyack was a block from the river, and when he was a teenager he sketched the boats and the nearby shipyards.

During the second half of the nineteenth century, harvesting and shipping ice to cities up and down the river and even farther afield was a booming business. African Americans, Italian and Irish immigrants, and both men and women worked at this harsh seasonal industry. From approximately 1840 to 1920 ice was harvested from the Hudson River, particularly north of Poughkeepsie. The ice near New York City was not used because it contained too much salt, which resisted freezing and melted quickly. Henry Golden Dearth's *Ice Boats on the Hudson* (fig. 1), painted in a cool tonal palette, depicts many boats docked at the height of the market for ice. Artists like Dearth who studied in Paris knew that by the 1870s both industrial and leisure activities along the Seine had become popular subjects of the French Impressionists and that exposure possibly influenced the greater acceptance of industrial scenes in American art.

Down river from Nyack is Hastings-on-Hudson, where Jasper Cropsey built a studio and home in 1885. In *The Hudson River at Hastings* (fig. 43), Cropsey shows the quarry works along with smoke-spewing chimneys from sugar

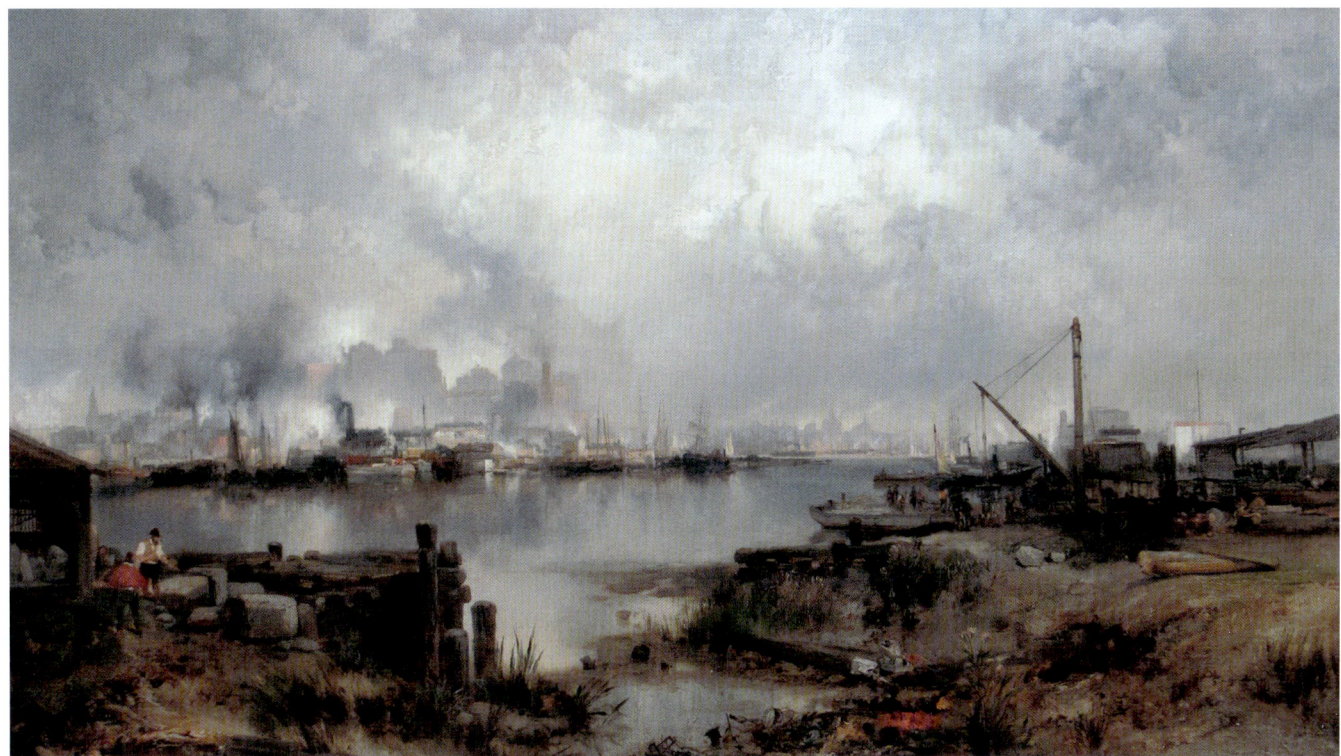

Fig 19. Thomas Moran (1837–1926). *Lower Manhattan from Communipaw, New Jersey*, 1880. Oil on canvas, 25¼ × 45¼ in. (64.1 × 114.9 cm). Washington County Museum of Fine Arts, Hagerstown, Maryland (A303.41.01)

refineries and paving-block factories where marble was prepared by skilled stonecutters and then loaded onto ships that carried it to New York and other cities. In the 1880s Hastings also produced millions of stone blocks that were used in Central and Prospect Parks and were shipped nationally and internationally. Cropsey mostly painted idyllic Hudson River scenes omitting factories; nevertheless, early on he recognized that in areas where the Lenape had once fished and collected oysters the atmosphere had become "soft and hazy when the air is filled with heat, dust, and gaseous exhalations."[5]

Thomas Moran, revered for his paintings of Yellowstone National Park, portrayed New York City from the docks on the opposite side of the river in *Lower Manhattan from Communipaw, New Jersey* (fig. 19). In the foreground he shows the devastation of the tidal marsh jutting into the Hudson River estuary that hosted vast oyster beds in the nineteenth century and the smog and smoke of the factories and buildings. Precursor of twentieth-century representations, Moran's painting boldly expresses industry's damage to the river and its shoreline, reminiscent of Thomas Cole's *The Course of Empire: Desolation* (1836; New-York Historical Society).

Changing times, tastes, aesthetics, and styles led artists to begin representing industry front and center. No longer omitted, veiled, or distanced, the railroad is portrayed as a major polluter by several artists in the

Fig. 20. George Bellows (1882–1925). *Rain on the River*, 1908. Oil on canvas, 32³⁄₈ × 38¼ in. (82.2 × 97.2 cm). RISD Museum, Providence, Rhode Island, Jesse Metcalf Fund (15.063)

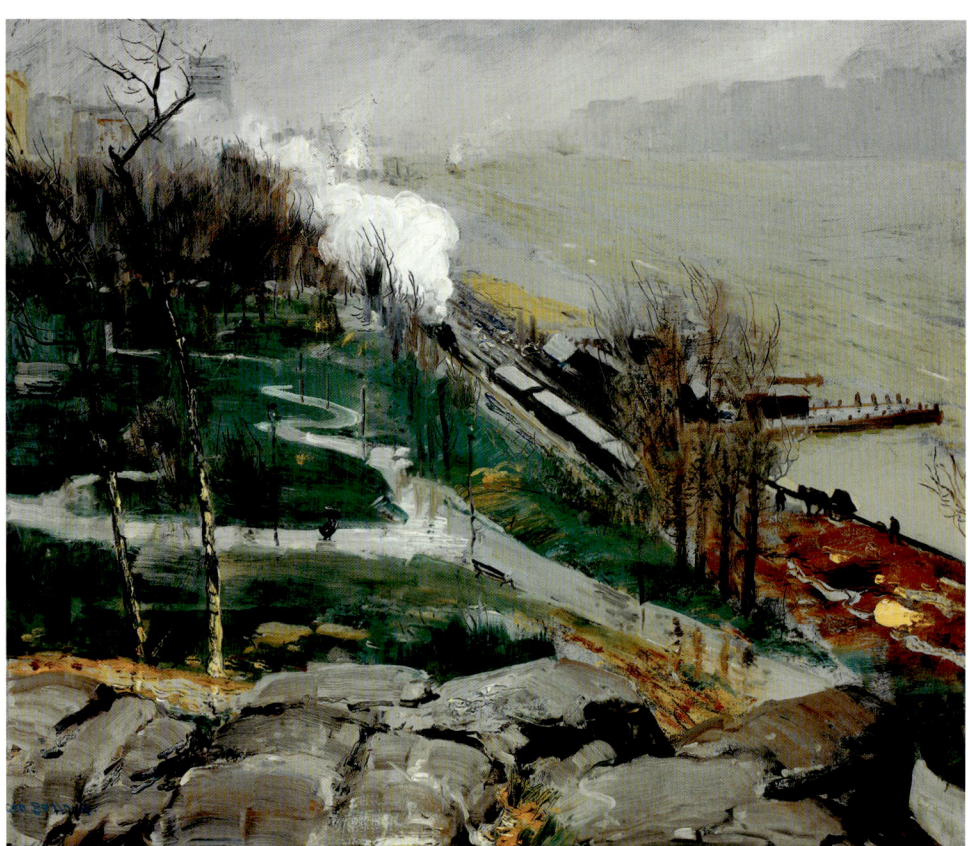

twentieth century. Between 1908 and 1912 realist painter George Bellows, known for his images of urban life, focused on the Hudson River in every season. With vigorous paint-laden brushstrokes, canvases such as *Rain on the River* (fig. 20) show the proximity of the railroad to the water. Originally Lenape land, the steep and rocky terrain was difficult to build on, so the railroad connecting Albany and New York was located close to the shoreline. Riverside Park, shown here, was still a work in progress. Between the coal emissions, garbage dumps, sewage, and squatters' shacks, its construction was as controversial as it was necessary.[6]

In *View of Manhattan from the Terminal Yards, Weehawken, New Jersey* (fig. 21) Abraham Leon Kroll foregrounds a rocky cliff to reveal the gritty west side of the waterfront, with its passenger trains, tracks, station, and multiple ferry slips. The smoke that winds into the hazy polluted atmosphere blurs the view of Manhattan in the distance. Two years later in *Freight Yards* (fig. 42) Gifford Reynolds Beal conveys industry as progressive, modern, and an economic necessity, one that made the river a transportation hub and New York City a center of commerce. The West 60th Street rail yards were on the Manhattan side of the river, opposite Kroll's view.

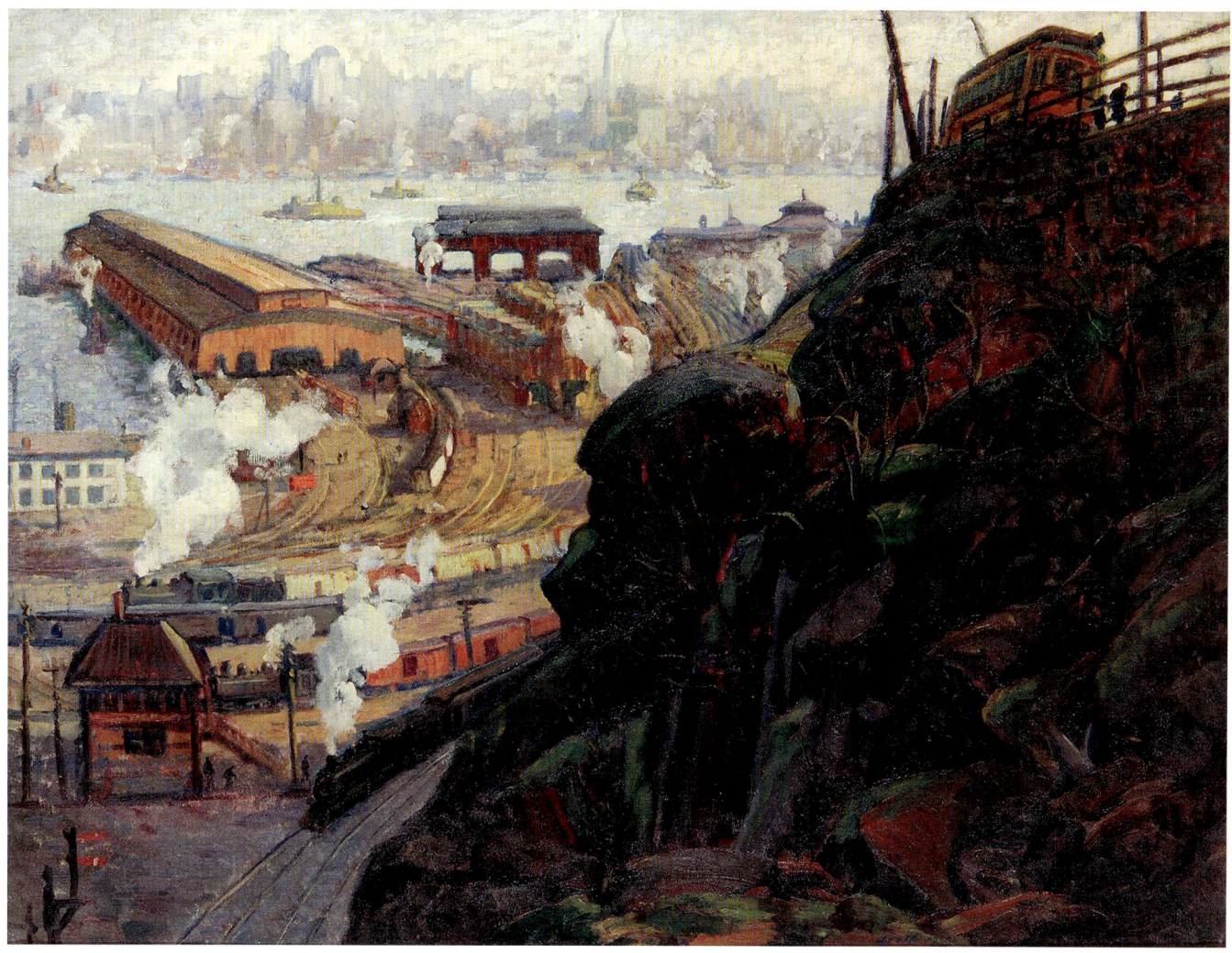

Revealing the general pollution of railroads, factories, and boats, Daniel Putnam Brinley's post-Impressionist *Hudson River View (Sugar Factory at Yonkers)* is modern in subject and style (fig. 22). This brightly colored, chaotic image, with plumes of smoke wafting diagonally to the sky from the horizontal lines of the railroad and paths below, is juxtaposed with the natural beauty of the river painted in paler tones. Sugar refining factories were all over New York, and many were located along the Hudson River for proximity to transportation. By 1900 New York State provided a base for the sugar industry, which flourished from the eighteenth century on, and also for the manufacture of cotton, wool, silk, and bricks, located along the Hudson River and Erie Canal corridor. Yonkers was a busy commercial port that housed train tracks, storage facilities, factory yards, and boat clubs.

Fig. 21. Abraham Leon Kroll, *View of Manhattan from the Terminal Yards, Weehawken, New Jersey*, 1913. Oil on canvas, 36 × 48 in. (91.4 × 121.9 in.). Montgomery Museum of Fine Arts, Alabama, Association Purchase: Art Acquisitions Fund and Gifts (2000.1)

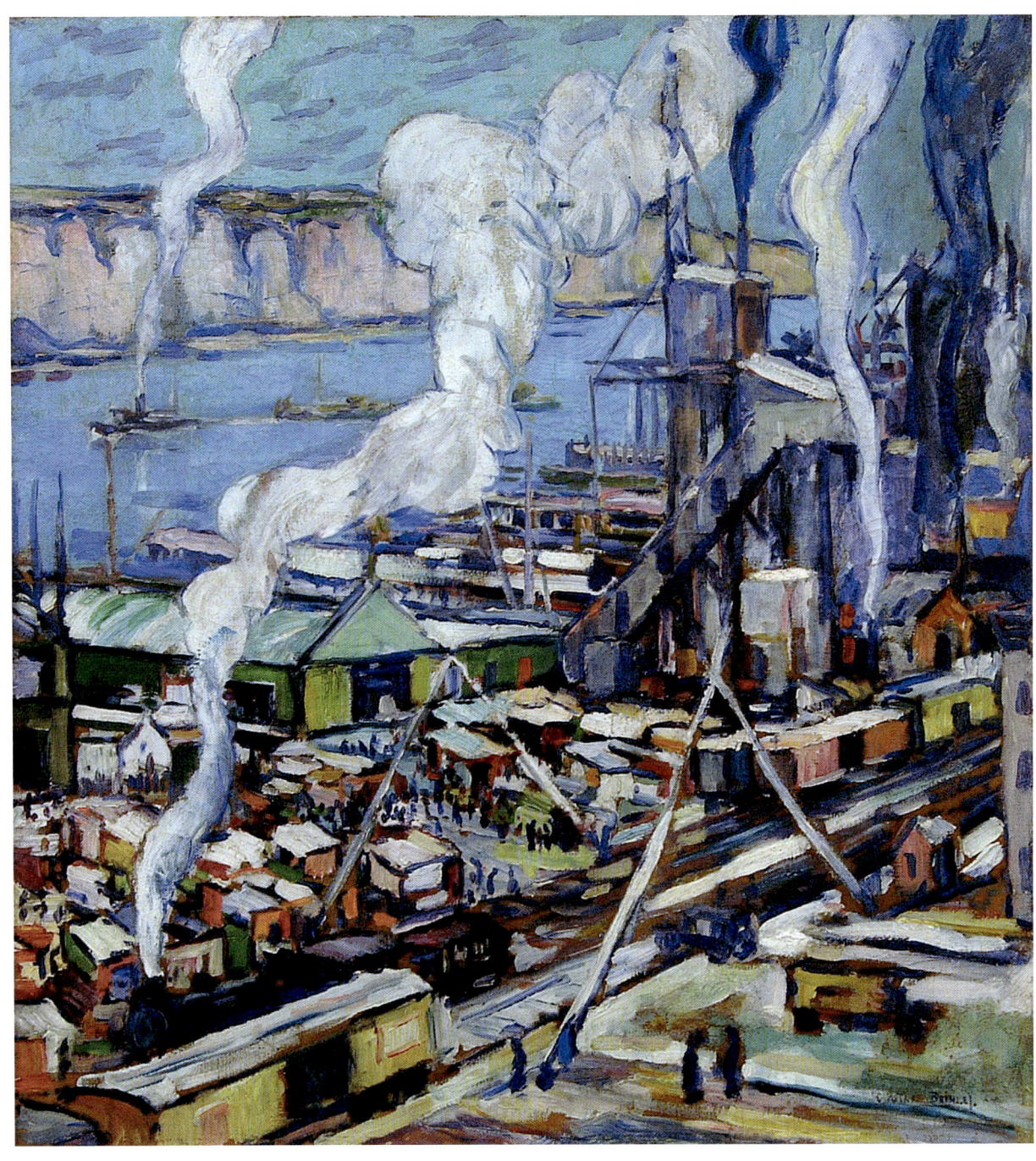

Fig. 22. Daniel Putnam Brinley (1879–1963). *Hudson River View (Sugar Factory at Yonkers)*, ca. 1915. Oil on canvas, 30 1/8 × 31 7/8 in. (76.5 × 80.9 cm). Hudson River Museum, Yonkers, Museum Purchase, 1995 (95.31)

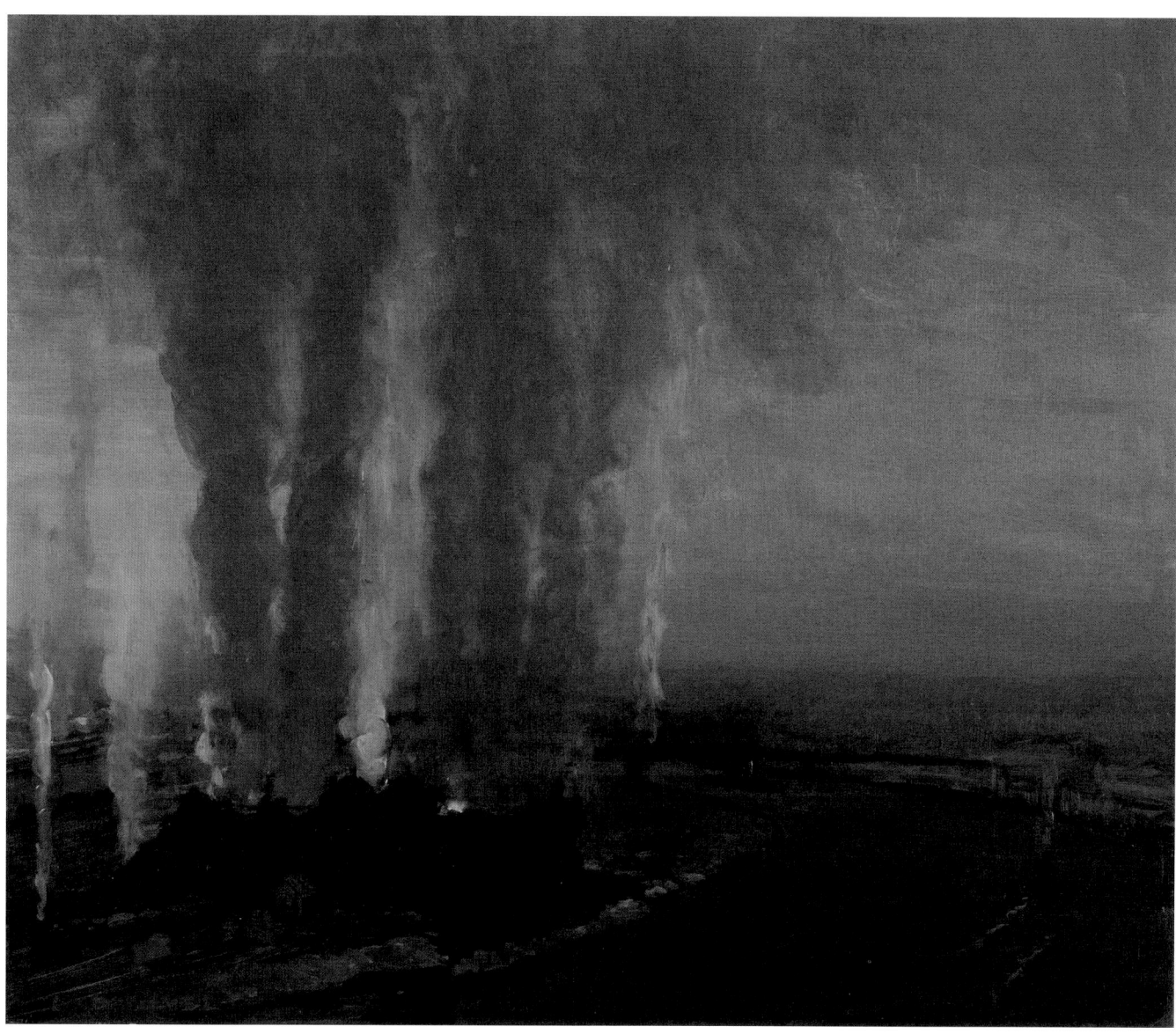

A prime example of railroad-related pollution is seen in George Benjamin Luks's *Roundhouse at High Bridge* (fig. 23), which boldly exhibits the contaminated emissions and noxious fumes from railroad buildings along the river in the Bronx. Luks was a member of the progressive group called The Eight and of the Ashcan School, whose artists focused on the gritty side of urban realism. With the advent of the railroad, American modernists produced a variety of subjects and styles, creating a dynamic look that broadcast the march of industry.

Another view from the Jersey side of the river is Henry Schnakenberg's *Edgewater, NJ* (cat. 66), depicting multiple industries but foregrounding the

Fig. 23. George Benjamin Luks, *Roundhouse at High Bridge*, 1909–10. Oil on canvas, 30¼ × 30½ in. (76.8 × 77.5 cm). Munson Williams Proctor Arts Institute, Utica, New York, Museum Purchase (50.17)

railroad tracks. The first decades of the twentieth century saw enormous industrial development in this area because of its access to the river, to railroads, to a labor force from the city, and to clients for its products. Cobblestones quarried from the cliffs, chemical factories, and the like are positioned between the greenery of the Palisades and the river, with the city in the distance blurred by pollution.

The decade of the Great Depression rejected traditional landscape conventions and proudly celebrated American modernism by promulgating industrial sites. Aaron Douglas, a key figure in the Harlem Renaissance during the 1920s and 1930s, painted *Inwood Power Plant* (fig. 24), previously titled *Power Plant in Harlem,* which demonstrates his sensitivity to the urban landscape. The title was changed because the Sherman Creek Generating Station portrayed is located at the intersection of 201st Street and the Harlem River in the Inwood neighborhood of Upper Manhattan rather than in Harlem. The station, built in 1913, was one of several power plants erected to meet New York City's increasing demand for electricity. Douglas's painting communicates the impact and importance of industry, the river echoing the colors of the station. Palmer Hayden, another Harlem Renaissance artist, created narrative scenes of New York's urban life and the rural South, specifically African American life, which he sometimes caricatured. Racism was a relevant topic in his art, but early in his career he painted images like *South Ferry* (fig. 47). Here, the turbulent waters of the Hudson are seen from the deteriorating piers.

During the early twentieth century artists once again focused on river transportation. Painter and printmaker Glenn O. Coleman's *Fort Lee Ferry* (cat. 18) shows both boats and the railroad, underscoring the connection between the two modes of transportation along the river. Travel was made accessible and convenient for people from all walks of life but added to the contamination of the Hudson and its environs. Ernest Fiene's lithograph *Hudson River Boat* (cat. 25) depicts one of the large steamers that plied the river along with ferries and tugboats and contrasts it with a nearby rowboat. In the background is the Bear Mountain Bridge, built in 1923–24, downriver from Woodstock, New York, where Fiene had a house.[7] Reginald Marsh's watercolor *Tugboat and New York City Skyline* (cat. 51) features the workhorse of river boats against the shoreline buildings as seen from New Jersey. Tugboats that guided larger vessels and barges were part of the scene that artists portrayed, reshaping modern American identity. Along with the railroad, the steamboats on the Hudson encouraged tourism and leisure activities in the Hudson Valley and farther north.

Artists found new sources of inspiration as they sought to capture the energy behind the skyscrapers, factories, and machinery, expressing

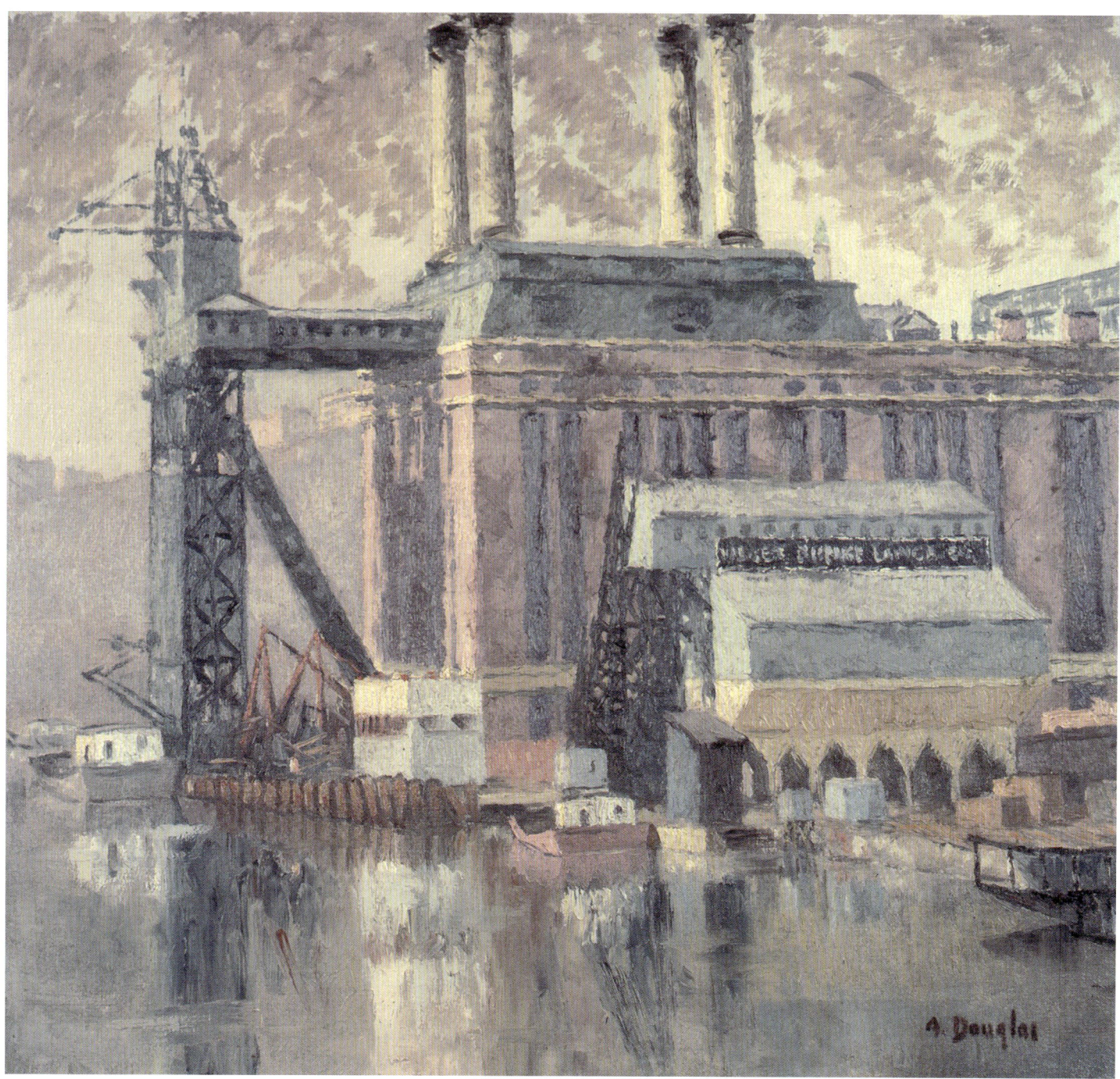

Fig. 24. Aaron Douglas (1899–1979). *Inwood Power Plant*, 1934. Oil on canvas, 20 × 22 in. (50.8 × 55.8 cm). Collection of the Hampton University Museum, Hampton, Virginia (67.112)

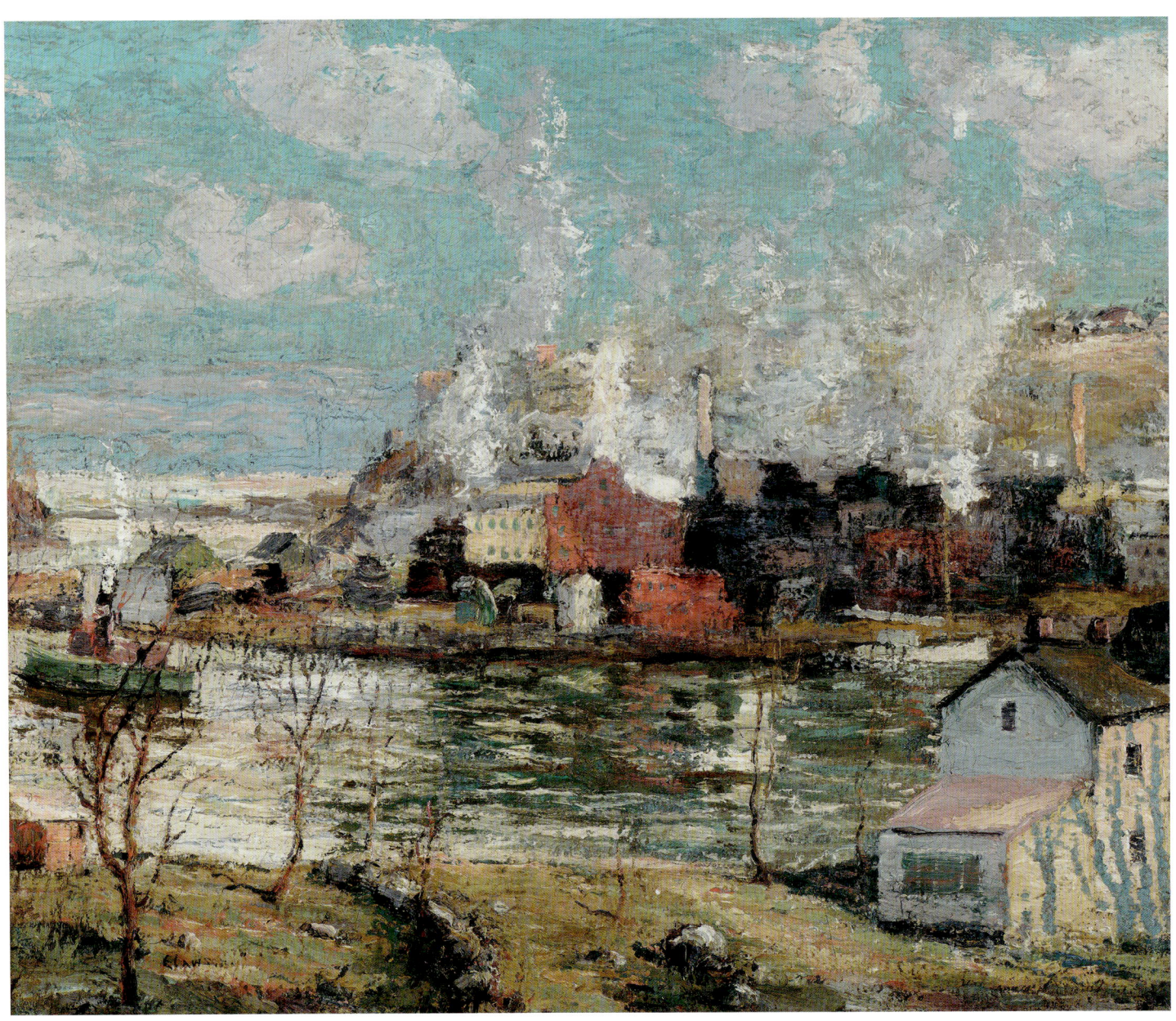

Fig. 25. Ernest Lawson (1873–1939). *Spuyten Duyvil Creek*, ca. 1914. Oil on linen, 25 × 30 in. (63.5 × 76.2 cm). Wichita Art Museum, Bequest of Glenn L. and Jayne Seydell Milburn (2017.20)

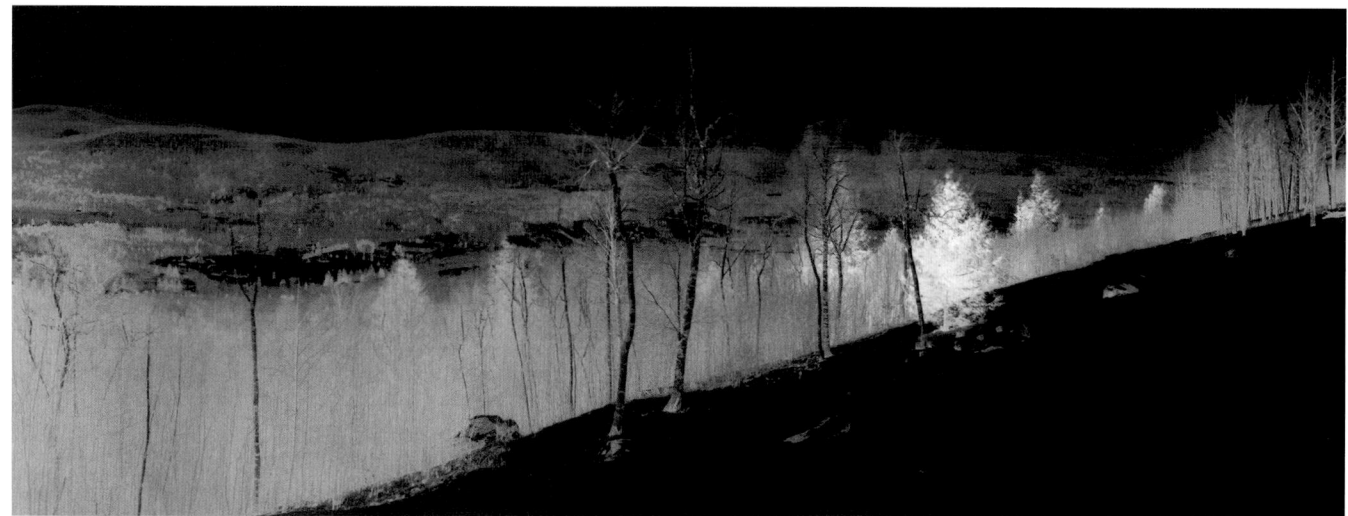

Fig. 26. Shi Guorui, *View of Catskill Mountains, New York, February 6–7*, 2019. Gelatin silver print, 45 × 115 in. (114.3 × 292.1 cm). Courtesy of the artist

modernism through industries set beside the timeless beauty of the Hudson. Several artists focused on the architectural developments along the river. Ernest Lawson, a member of The Eight, was the only landscapist among them. He documented the urban frontier, marking the encroachment of city buildings on what had been rural open land. *Spuyten Duyvil Creek* (fig. 25) displays a handling of paint and discreet color influenced by French and American Impressionism. For the next two decades Lawson, who moved to Washington Heights in Upper Manhattan in 1898, focused on the rocky inclines of this remote, underpopulated edge of the metropolis that had once been rich in fish and wildlife.[8] Spuyten Duyvil, in the southwest corner of the Bronx, directly north of Inwood in Manhattan, is situated on a bluff overlooking the Hudson River on the west and Spuyten Duyvil Creek to the south. This short tidal estuary connecting the Hudson River to the Harlem River separates the island of Manhattan from the Bronx. The Dutch name is translated variously as spinning or spitting devil, so-called for its strong currents.

Spuyten Duyvil, First View of the Hudson (cat. 39), a picture of a rural road along the river, is by William Henry Jackson, best known for his images of the American West. In the exhibition it is representative of the increasing use of photography to document the river. Photographer Shi Guorui, using a camera obscura, combines vintage and contemporary techniques to create expanded images of sites that Thomas Cole painted. The velvety black-and-white *View of Catskill Mountains, New York, February 6–7* (fig. 26) can be seen as an abstraction or extraction of light and time.[9]

The 1930s and 1940s brought about a new kind of realism, diversity of styles, and flexibility to accommodate America's changing environment and

OMISSIONS FROM THE CANON

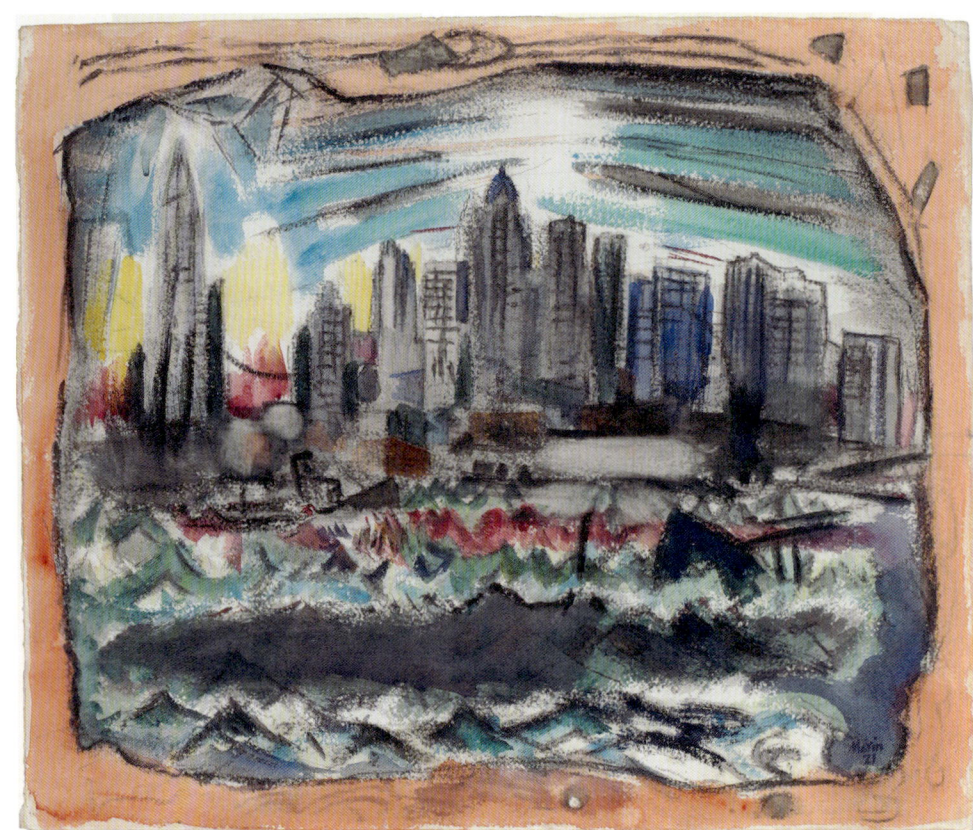

Fig. 27. John Marin (1870–1953). *Lower Manhattan from River, No. 1*, 1921. Watercolor, charcoal, graphite on paper, 21⅞ × 26½ in. (55.6 × 67.3 cm). The Metropolitan Museum of Art, New York, Alfred Stieglitz Collection, 1949 (49.70.122)

experiences. In the watercolor *Lower Manhattan from River, No. 1* (fig. 27) John Marin shows how buildings impacted the shoreline, as seen from the Jersey side. Marin conveys modern life during this period of great optimism about the country's progress—proclaiming new forms of transportation, communication, and industrialization—without indicating the devastation to the environment.

In *New York City Nocturne* (cat. 28) Emil Ganso uses color and brushstroke to convey the energy of the city, its traffic and its lights emanating from skyscrapers that reflect on the river. This concentration on the city's architectural achievements was continued by Yvonne Jacquette's ariel views, as seen in her woodcut *Hudson River Diptych* (fig. 52). Both Ganso and Jacquette celebrate the city's unique skyline without processing its impact on the Hudson waterfront.

The negative effects of industrialization on the river and the natural environment were increasingly recognized in the mid-twentieth century. By including works that are relatively unknown, subjects not often tackled by earlier artists and rarely published in the literature, *Shifting Shorelines* reveals how artists responded to industrial and ecological issues, how aesthetic values evolved, and, most significantly, how we extract a new narrative pertinent to today's economic, cultural, and social concerns. Just as the shoreline shifted, so did the interest in, and acceptance of, artistic images of industry and ecology.

Notes
1. Thomas Cole, "Essay on American Scenery," *American Monthly Magazine* 1 (January 1836); reprint, Thomas Cole National Historic Site, Catskill, N.Y., 2018, 15–16.
2. For more about the Catskill Creek series, see H. Daniel Peck, *Thomas Cole's Refrain: The Paintings of Catskill Creek* (Ithaca, New York: Three Hills, An imprint of Cornell University Press, 2019).
3. William Zeisel, *Angling on a Changing Estuary: The Hudson River, 1609–1995*. Report to the Hudson River Foundation, New York, 1995.
4. Despite barriers of equity and education, there were many successful and prolific women artists who produced works that appealed to a rising middle class yet were erased from history by gender bias. After her husband died in 1856, Beers supported herself by the sale of paintings supplemented by group sketching trips. Some people attributed Beers's success to her siblings, painters William and James Hart, but she insisted that it was due "to sheer hard work and persistent labor." See Katherine Manthorne, *Restless Enterprise: The Art and Life of Eliza Pratt Greaatorex* (Oakland: University of California Press, 2020), 66.
5. Jasper Francis Cropsey, "Up among the Clouds," *Crayon* 2, no. 6 (August 8, 1855): 337.
6. Some people objected to the land condemnation and the construction noise of Riverside Park, designed by Frederick Law Olmstead.
7. Fiene painted two versions of this image in oil without the rowboat: Whitney Museum of American Art, New York; and D. Wigmore Fine Art, New York.
8. Ross Barrett, "Speculations in Paint: Ernest Lawson and the Urbanization of New York," *Winterthur Portfolio* 42, no. 1 (2016): 1–25.
9. See Kate Menconeri, *Shi Guorui: Ab/Sense-Pre/Sense* (Catskill, N.Y.: Thomas Cole National Historic Site, 2019).

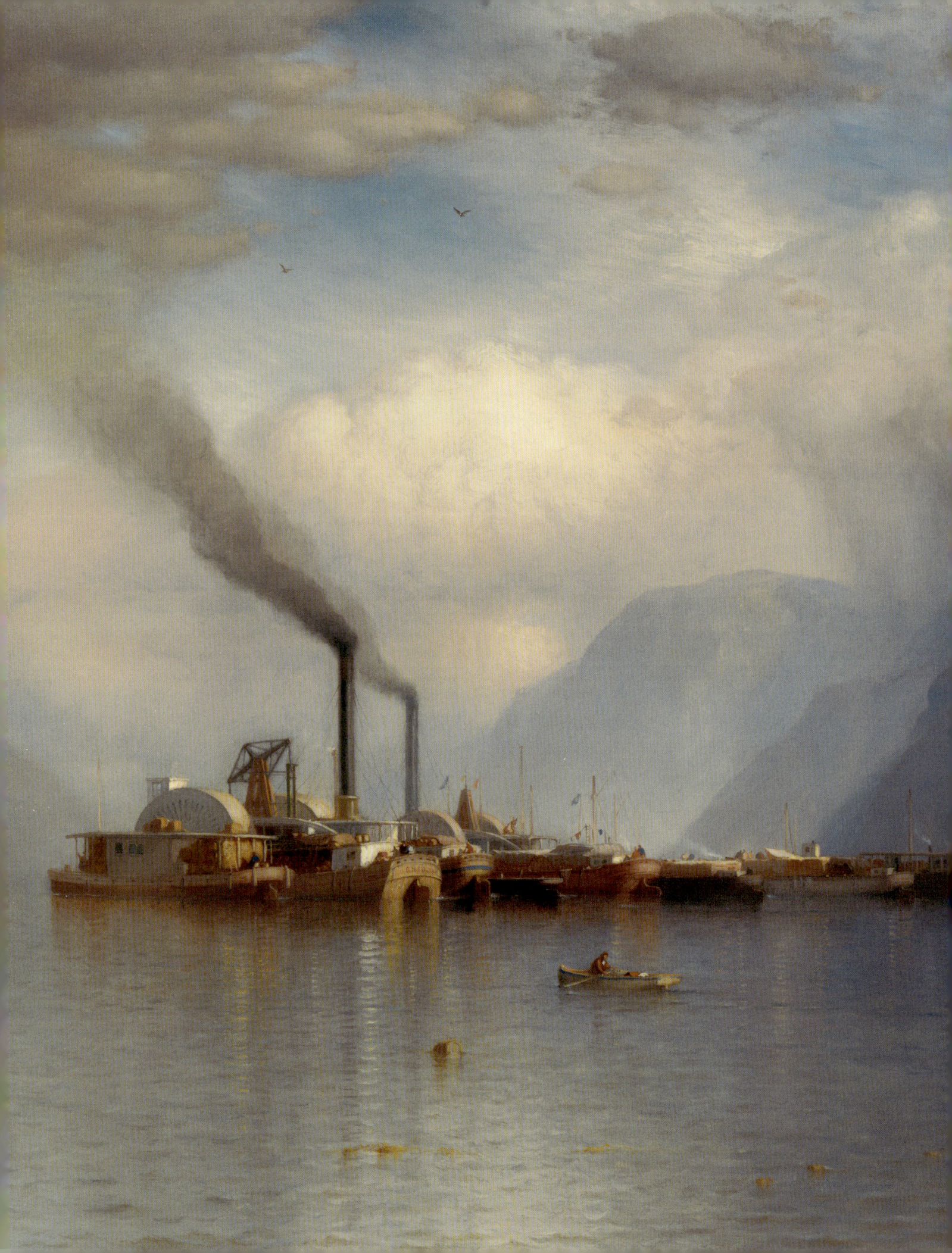

Industrial Harmony

Connecting Industry along the Hudson River Railroad

Victoria Horrocks

The 1867 review of Samuel Colman's painting *Tow Boats on the Hudson* in the *American Art Journal* extols the work's "rich color and great truthfulness."[1] Now one of Colman's most recognized landscapes of the Hudson River, the painting—later renamed *Storm King on the Hudson* (fig. 28)—depicts the major water route populated by steamships, sailboats, and rowboats.[2] Insofar as Colman's picture is an accurate representation of industry along the Hudson in 1866, the painting is indeed "truthful." Ships at left, varying in size and shape, appear to be steamers, often used for cargo or passenger travel and occasionally for towing barges.[3] Men in a rowboat in the foreground are casting a fishing net, while sailboats in the distance suggest both leisure activities and sloops—the predecessor to the steamer for shipping cargo.[4] The flourishing of industry portrayed in *Storm King on the Hudson* was undeniably due to a crucial development unfolding concurrently with Colman's scene: the construction of the Hudson River Railroad.

At second glance, the railroad is immanent in Colman's painting due to its role as an essential conduit for the many industries emerging in the Hudson Valley at the time. Where a thick stream of black smoke rises from ships and darkens the otherwise white clouds evocative of the Hudson's natural beauty, so, too, was the shoreline of the river reconfigured by the railroad and the proliferating industries it supported. The water route in *Storm King on the Hudson* is located near the railhead for southbound trains headed toward New York City.[5] It is safe to assume, then, that the steamers and freight ships in the painting could be offloading their cargo and leisure travelers onto nearby trains.

On August 1, 1846—twenty years prior to the completion of Colman's painting—authorities of the Hudson River Railroad convened in Poughkeepsie to discuss the need for a robust railway system. Despite the financial drain of construction (initial projections in 1842 estimated that the project would require roughly three million dollars in capital), the chairman, David B. Lent, argued vehemently that a railroad was necessary to enact competitive commerce in the State of New York.[6] Believing its construction to be a "defensive measure," Lent warned that burgeoning railway systems

Samuel Colman. *Storm King on the Hudson* (detail), p. 60

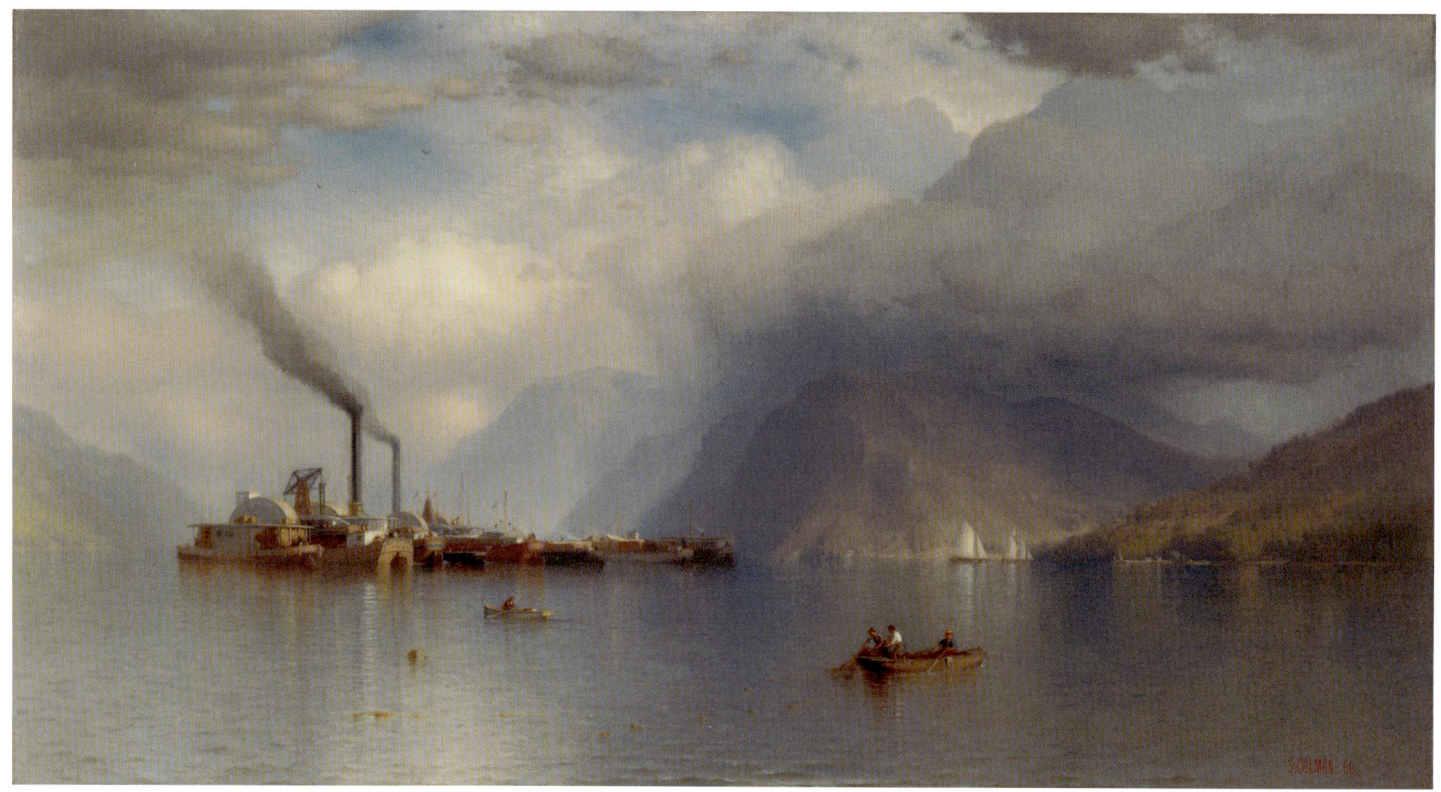

Fig. 28. Samuel Colman (1832–1920). *Storm King on the Hudson*, 1866. Oil on canvas, 32 1/8 × 59 7/8 in. (81.6 × 152.0 cm). Smithsonian American Art Museum, Gift of John Gellatly (1929.6.20)

in Pennsylvania and Massachusetts were enabling these neighboring states to lead the country in trade and commerce. In a grandiose turn of phrase, he writes, "It is time now to look to another and more powerful engine of internal improvement, and to the third competitor, that has not hitherto taken the field. *Railroads* have been preparing the way for another revolution in trade."[7] Throughout the Northeast during this period, railroads were superseding canals for trade, and while the Erie Canal had been an exception to this economic tendency, Lent believed that river passage would eventually become ineffectual and that a railroad would be essential "in securing quester trade to this state and the city of New York."[8]

Lent astutely predicted the need for a railway system throughout the Hudson Valley, but the ways in which the railroad supported industry far exceeded his predictions. Most notably, it enabled more efficient transportation of the Hudson Valley's wealth of natural resources: iron ore, limestone, basalt, coal, charcoal, and water.[9] The production of iron and the extraction of basalt fostered the construction of sidewalks, highways, and more. Passenger travel also became more accessible and thus increasingly popular: the West Point Foundry, for example, drew steamers past Storm King Mountain, and neighboring towns inspired tourist travel upriver to hotels and other attractions.[10] Running along the Hudson's shores, rail-related commerce complemented the existing endeavors happening on the river.

Storm King on the Hudson elegantly portrays the significant industrial turn during the second half of the nineteenth century, both in what is rendered by Colman's brush and in what the scene allows observers to infer about the railroad's role in supporting the water route at Storm King Mountain and beyond. The "truthfulness" celebrated in the painting's 1867 review is not merely a note on historical accuracy but further situates the work in the larger genre of landscape painting pioneered by the Hudson River School. Works by these artists, such as Thomas Cole, Frederic Edwin Church, Susie Barstow, and others, critically employ images of the natural beauty of the Hudson Highlands "as an arena of symbolic action, a quasi-utopian endeavor that helped to order culturally a space inherently open-ended and unstable."[11] According to Angela Miller, landscape paintings like Colman's conveyed an "uncomplicated harmony" that communicated a "national unity, pride of place, and unique identity distinct from that of Europe."[12] Amid the instability of changing modes of transportation, developing infrastructure, and new technologies, *Storm King on the Hudson* is a portrait of American economic growth. The "quasi-utopian endeavor" captured in the painting conveys not only a harmony between the natural landscape and industrial development but also a harmony between the distinct industries ultimately connected by the Hudson River Railroad.

Notes
1. Paletta, "Art Matters," *American Art Journal* 6, no. 5 (January 5, 1867), 165, cited in John K. Howat, et al., *American Paradise: The World of the Hudson River School* (New York: Metropolitan Museum of Art, 1987), 304.
2. Howat et al., *American Paradise*, 303.
3. Hudson River Maritime Museum, "The Hudson River Sloop," *Hudson River Maritime Museum*, November 6, 2016; https://www.hrmm.org/history-blog/the-hudson-river-sloop.
4. Ibid.
5. "Storm King on the Hudson," *Smithsonian American Art Museum*, https://americanart.si.edu/artwork/storm-king-hudson-5183.
6. Frances Dunwell, *The Hudson River Highlands* (New York: Columbia University Press, 2008), 132.
7. William H. Grant and Jervis John Bloomfeld, "The Hudson River Railroad. Observations on the western trade, and its influence upon the growth and prosperity of the cities of New York, Boston and Philadelphia through several competing lines of communication and the Hudson River Railroad (Poughkeepsie: Journal & Eagle Printing Establishment, 1846), 7, 12. Emphasis in original.
8. Ibid., 24.
9. Dunwell, *Hudson River Highlands*, 111–12.
10. Ibid., 109.
11. Angela Miller, *Empire of the Eye* (Ithaca: Cornell University Press, 1993), 12.
12. Ibid.

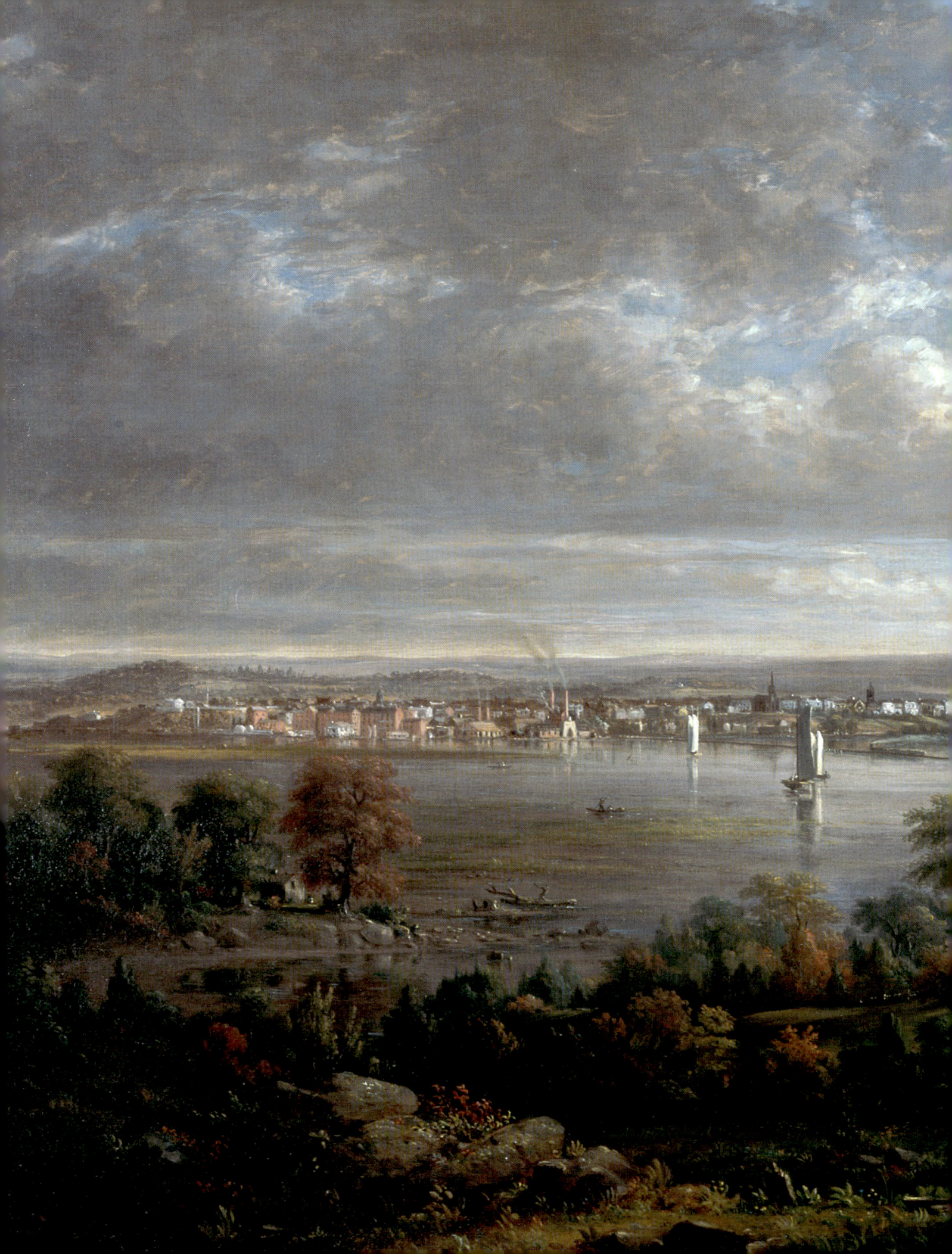

Hudson Marshes

Archives of Biodiversity, Climate Change, and Human Impact

Dorothy M. Peteet

Shoreline art conveys the history of change through centuries, but its most significant impact is the emotion it engenders. Linking science and art enriches the experience of both as well as our understanding of the environment and our changing roles within it. The Hudson shoreline includes wetlands that link the river water to the uplands. These wetlands are critical for purifying water; providing habitat for fish, birds, amphibians, and mammals; and protecting the coastline from storms. The dark brown peat buried below the waves is anoxic and cold, thus preserving deep layers of archival knowledge and carbon stocks that are valuable as our forests. Discoveries from Piermont Marsh are one example of paleoenvironmental change contained in these wetlands and reflected in art.

Melting of the last Ice Age glacier about sixteen thousand years ago left deep deposits of clay beneath the ice, lining the fjord's shoreline from glacial Lake Champlain southward to the tip of Manhattan, where it meets the North Atlantic Ocean.[1] Bringing contributions of sand, silt, clay, and fresh water, at least eighty tributaries entered the Hudson River from Troy southward,[2] providing the perfect substrate for wetland vegetation to grow at these junctures. Tidal marshes then formed in this estuary, slowly building up over millennia, the oldest spanning more than six thousand years.[3] We extracted sediment cores of this valuable fossil peat as deep as forty feet atop the basal clays on both sides of the Hudson. Marsh records including Ramshorn Swamp near Catskill, Tivoli Bays near Red Hook, Haverstraw Marsh, Iona Island Marsh, Piermont Marsh, and Jamaica Bay provide us with the paleo history of forests, climate shifts, marsh biodiversity, and human impact (fig. 29).

The ubiquitous shoreline gray clay—tiny fragments of rock ground up by the glacier—was first utilized by Native Americans to craft vessels for food, as demonstrated by pottery in the 2023 *Indigenous Woodsplint Baskets* exhibition in New Paltz. Women in these Native American communities also wove baskets using red maple (*Acer rubrum*) and black ash (*Fraxinus nigra*) from the wetland shoreline swamps as illustrated by the Mohican split ash basket (fig. 30). The early April leafless mountains of the highlands, with the calm shoreline featuring lime-green freshwater marshes, are illustrated

Henry Ary. *View of the Hudson* (detail), p. 69

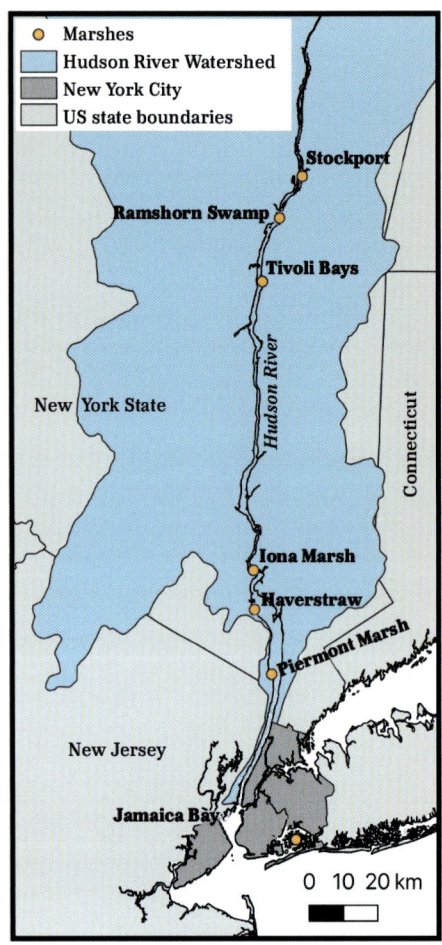

Fig. 29. Map of Hudson River sediment core extractions. Lamont-Doherty Earth Observatory, Palisades, New York

Fig 30. Algonquian, Mohican. Plaited basket, n.d. Split ash and ink, 10½ × 16 × 13 in. (26.7 × 40.6 × 33 cm). New York State Museum, Albany (E-29539A-B)

in Worthington Whittredge's *Shad Fishing on the Hudson* (fig. 16). Lenape boats perfectly adapted for marshes are beautifully illustrated by the 1651 print of New Amsterdam, reflecting nearby Jamaica Bay (fig. 3).

Just as they do today, these marshes filtered water, served as nurseries for fish, and provided food and cover for the young. Resident fish and anadromous species such as alewives, striped bass, sturgeon, and shad in turn nourished both people and shoreline fauna such as bear, raccoons, and bald eagles. Clams, striped bass remains, and sturgeon scutes found in the Huyler rock shelter overlooking the Hudson at Hyde Park, along with net sinkers from both the rock shelter and a site on the river's east bank at Newton Hook (fig. 31) attest to the abundant resources that the river and its margins provided to the Esopus Native Americans upriver and to the Lenape downriver in Manhattan.

Great blue herons, wood ducks, green herons, and Virginia rails continue to make extensive use of these marshes throughout the year,[4] and migrating birds alight in fall and spring. The northernmost tidal swamp, Ramshorn-Livingston, located near Catskill, is one of the largest (1,300 acres) remaining freshwater tidal marshes in the Hudson.[5] The core reveals that the swamp originally included alder (*Alnus*), a nitrogen-fixing plant that is used to make a brown dye, and many native sedges[6] such as sawgrass

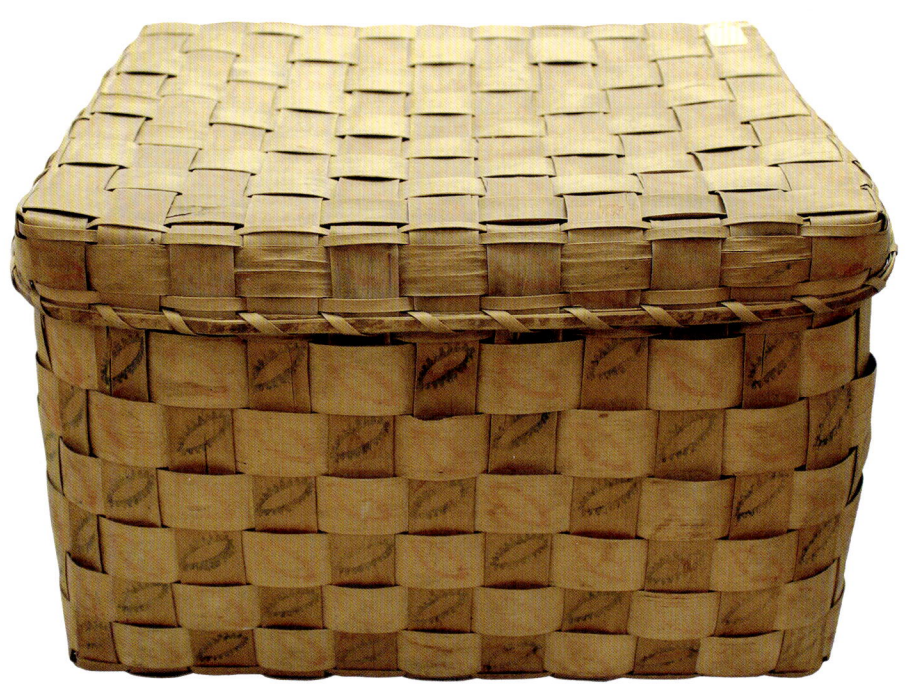

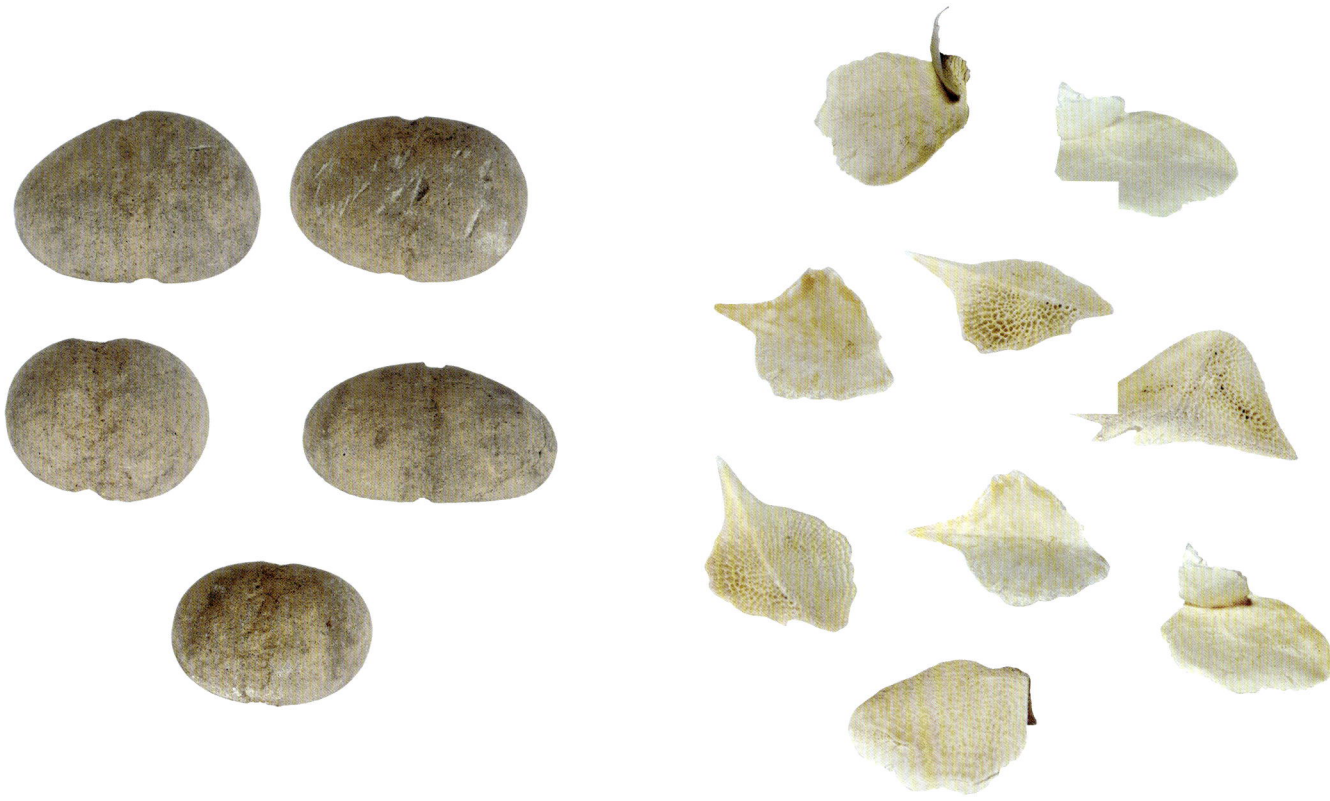

(*Cladium*) and bulrush (*Schoenoplectus*), used for making baskets and even possibly for lining wigwam roofs. Also present were seeds of sweet flag (*Acorus americanus*), the second most utilized plant in North America (216 uses) by Native Americans,[7] which has now disappeared in most of the Hudson marshes. Haverstraw marsh, however, still has a small population that remains but is slowly disappearing[8] along with the wild rice (*Zizania*) that was sustaining for the Lenape. Across the river on the eastern side and a little farther south, pollen from Tivoli Bay marsh near Red Hook reveals that the upland forest composition about a millennium ago primarily comprised oak (*Quercus*), hickory (*Carya*), and pine (*Pinus*), with some elements of spruce (*Picea*), hemlock (*Tsuga*), and birch (*Betula*).[9] This rich upland forest environment changed after the 1681 purchase of Cruger Island,[10] when Dutch and English in the eighteenth century logged the oak and pine for structures and fuel, farmed the fields, and stripped the hemlock for tanneries. Julie Beers's *Cows in a Landscape* (fig. 17) depicts pastures surrounded by forest remnants, cows scattered across the fields, and stone walls, indicating significant disturbance of the landscape, evidenced by higher mineral runoff and increased nitrogen pollution in the Hudson shoreline marshes downstream.

Fig. 31 (left). Mohican, 5500 BCE–1609 CE. Cache of five pebble net sinkers from the east bank of the Mahicannituck (Hudson River) at Newton Hook. Collection of Stockbridge-Munsee Band of Mohicans

Fig. 32 (right). Male Atlantic sturgeon (*Acipenser oxyrinchus*) scutes from a dying fish caught during a research initiative, June 1, 2022, at Mahicannituck (Hudson River) near Hyde Park, New York. Collection of Stockbridge-Munsee Band of Mohicans

The early nineteenth century saw vast hydrological changes to the shoreline. New York State constructed dikes to block offside channels and spur dikes to narrow the channels and increase velocities; these efforts were expanded by the United States Army Corps of Engineers, which dredged and constructed larger longitudinal dikes along the Hudson in the 1880s.[11] Continuous coastline renovation was caused by the early brickmakers along the shorelines from Albany to Haverstraw who mined the rich blue clay deposits that were more than one hundred feet thick, denuded the forest for fuel for baking the bricks, and then shipped the bricks on barges for the towering New York City landscape.[12] Excavations of clay seen in Marie-François-Régis Gignoux's *Haverstraw on the Hudson* (fig. 18) are sweeping from one side of the bay to the other, illustrative of the vast scope of the riverbank extraction and utilization.

Between 1800 and 1972 shorelines and wetlands were extensively altered, eliminated, or relocated along the 152-mile estuary between Troy and Catskill, and more than one-third of the surface of the river in this reach—some 3,300 acres—was filled with sediment from the federal navigation channel.[13] Seawalls, jetties, bulkheads, and railroads all define shoreline hardening, which results in losses of nutrient cycling[14] and biodiversity.[15]

The Tivoli Bays macrofossil record depicts rich fern deposits along with the sedge marsh composition (*Carex aquatalis*), violets (*Viola*), and aquatic water nymphs (*Naias flexilis*). With the arrival of Europeans, everything changed, both upriver and downriver, as recorded by the marshes. Instead of the abundant diversity of marsh plants, monocultures of plants began to establish themselves due to higher nutrient (nitrogen and phosphorus pollution) concentrations as greater populations of animal and human waste flowed into the river. A hybrid cattail (*T. glauca*) of *T. angustifolia* × *T. latifolia* (the native wider leaf cattail) began to grow vigorously in the marshes, replacing the native sedges. Lisa Sanditz's *Tivoli Bay* (fig. 33) reveals the severe interruption of river water by the railroad that formed these shallow bays, cutting off most of their twice daily tidal flow and resulting in more stability and cattail dominance. With increased sediment load and more stasis instead of a pulsing regime, the shoreline topography became more homogenous instead of heterogenous.[16] This process of water stabilization, coupled with urbanization with higher nutrients, has fostered expansion of *T. glauca* throughout the United States, resulting in enormous biodiversity decline in wetlands.[17]

Paleoecological records from Iona Marsh near the Bear Mountain Bridge[18] and Piermont farther south[19] show the dominance of the same native sedges as in Ramshorn Swamp. The increase of ragweed (*Ambrosia*), as trees (pine, hemlock, oak, hickory) were cut in the early eighteenth century for farming,

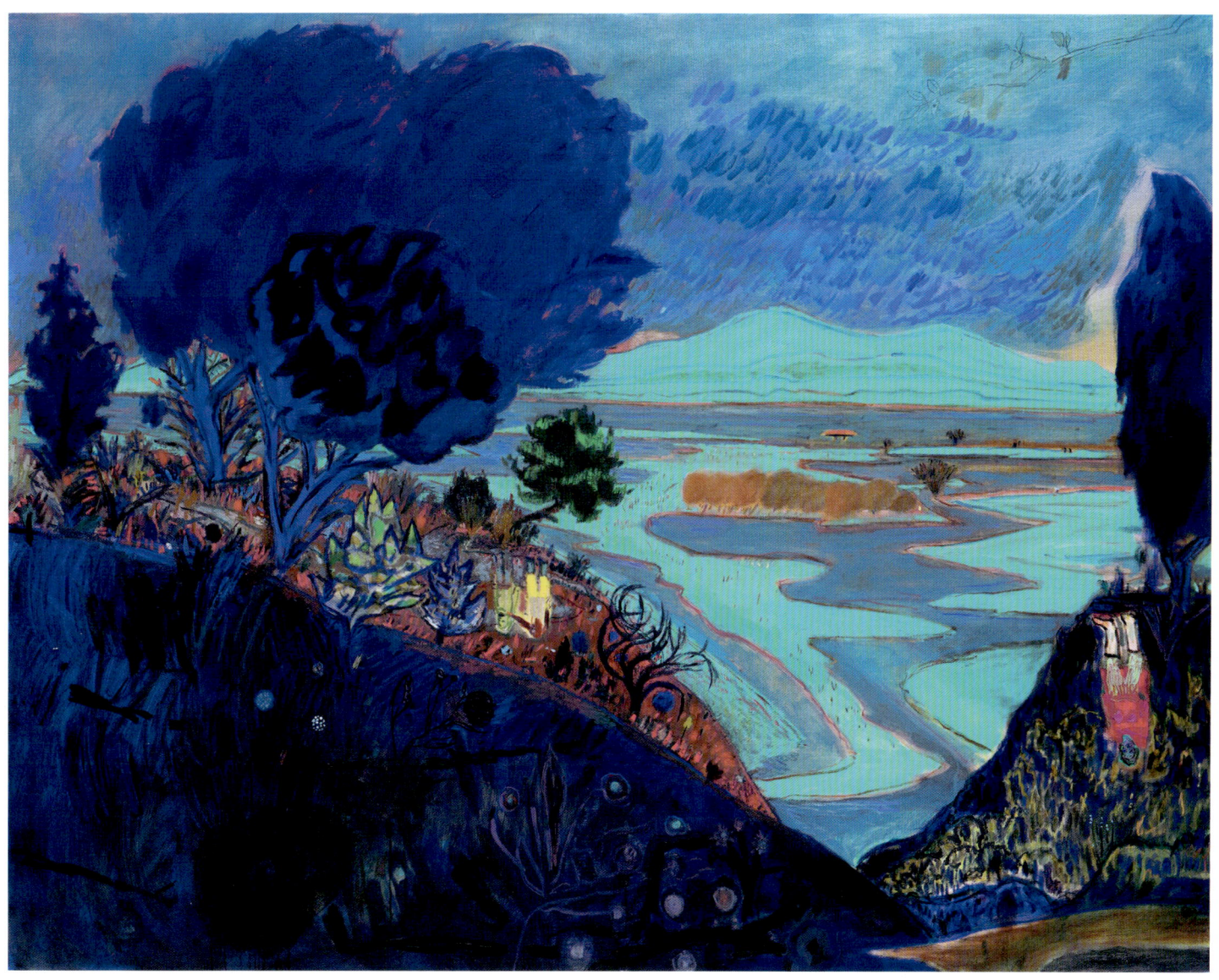

Fig. 33. Lisa Sanditz (b. 1973). *Tivoli Bay*, 2016. Oil on canvas, 54 × 70 in. (137.2 × 177.8 cm). Collection of the artist, courtesy Huxley Parlour Gallery, London

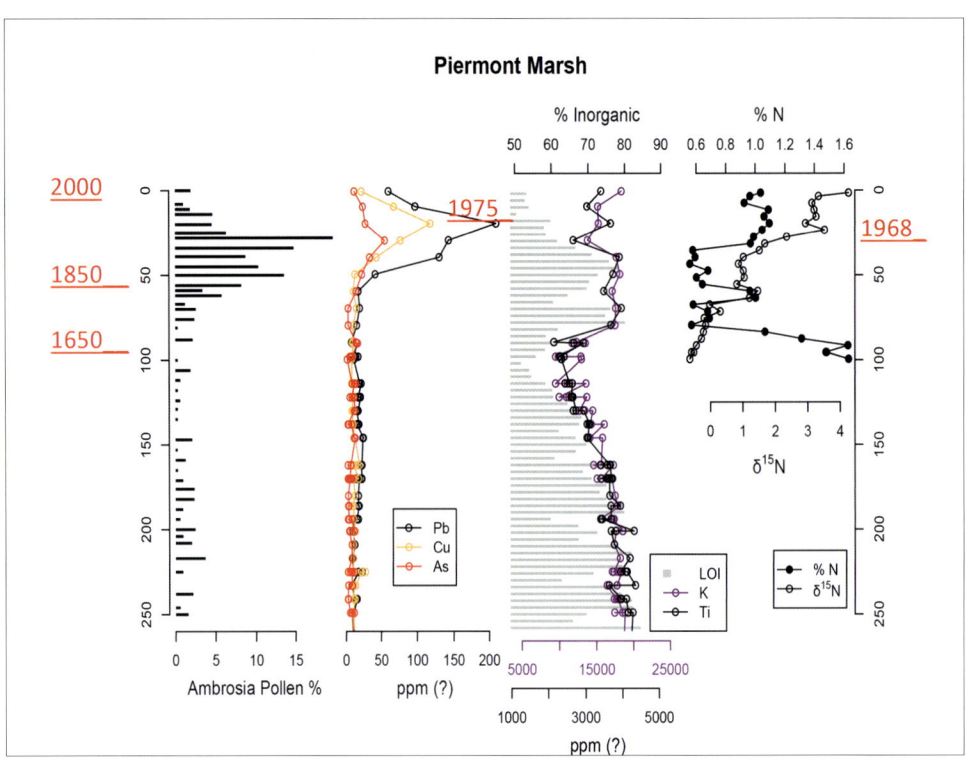

Fig. 34. Piermont Marsh sediment core data showing the rise of *Ambrosia* (ragweed), indicating deforestation, and the rise of heavy metals (Pb-lead, Cu-copper, and As-arsenic), indicating industrial pollution. The data also reflect the percentage of inorganic (mineral), K (potassium), and Ti (titanium) from upland erosion and the rise of nitrogen from agriculture, sewage outflows, power plants, and vehicle emissions.

fuel, and roads, and the large influx of upland inorganic sediment (fig. 34) affected the wetlands along the shore. Higher nutrients and stabilized water levels encouraged the spread of cattails and later common reed (*Phragmites*) from Europe.[20] Henry Ary's *View of the Hudson* (fig. 35) shows the dense level of development and hardening of the shoreline as smokestacks pollute the air, Hudson waters appear murky, and uplands are lacking in dense forests.

The rise in ragweed is carbon-14 dated with plant macrofossils in the core to about 1650 CE, mirroring the decline in trees observed in sediment cores from marshes up and down the river. Prior levels of charcoal and dominance of pine and hickory document the Medieval Warm Period (ca. 800–ca. 1250 CE) that the Hudson experienced, linking climate change from North America to Europe.[21] The sediment core stratigraphy further illustrates in detail the heavy metals of lead, copper, and zinc pollutants emanating from the smokestacks of the iron, brick, and coal-burning factories, steamships, and railroads along the river. Paintings in the mid-nineteenth century depicting this air pollution, which was eventually deposited in the marshes, include Johann Herman Carmiencke's *Poughkeepsie Iron Works (Bech's Furnace)* (fig. 36), with at least six chimneys belching colored smoke. Similarly, Frederic Edwin Church's *Hudson River with Factory by Moonlight* (fig. 9) reveals the twenty-four-hour year-round nature of many emissions, ranging from the coal-burning cement factories upriver such as those of the

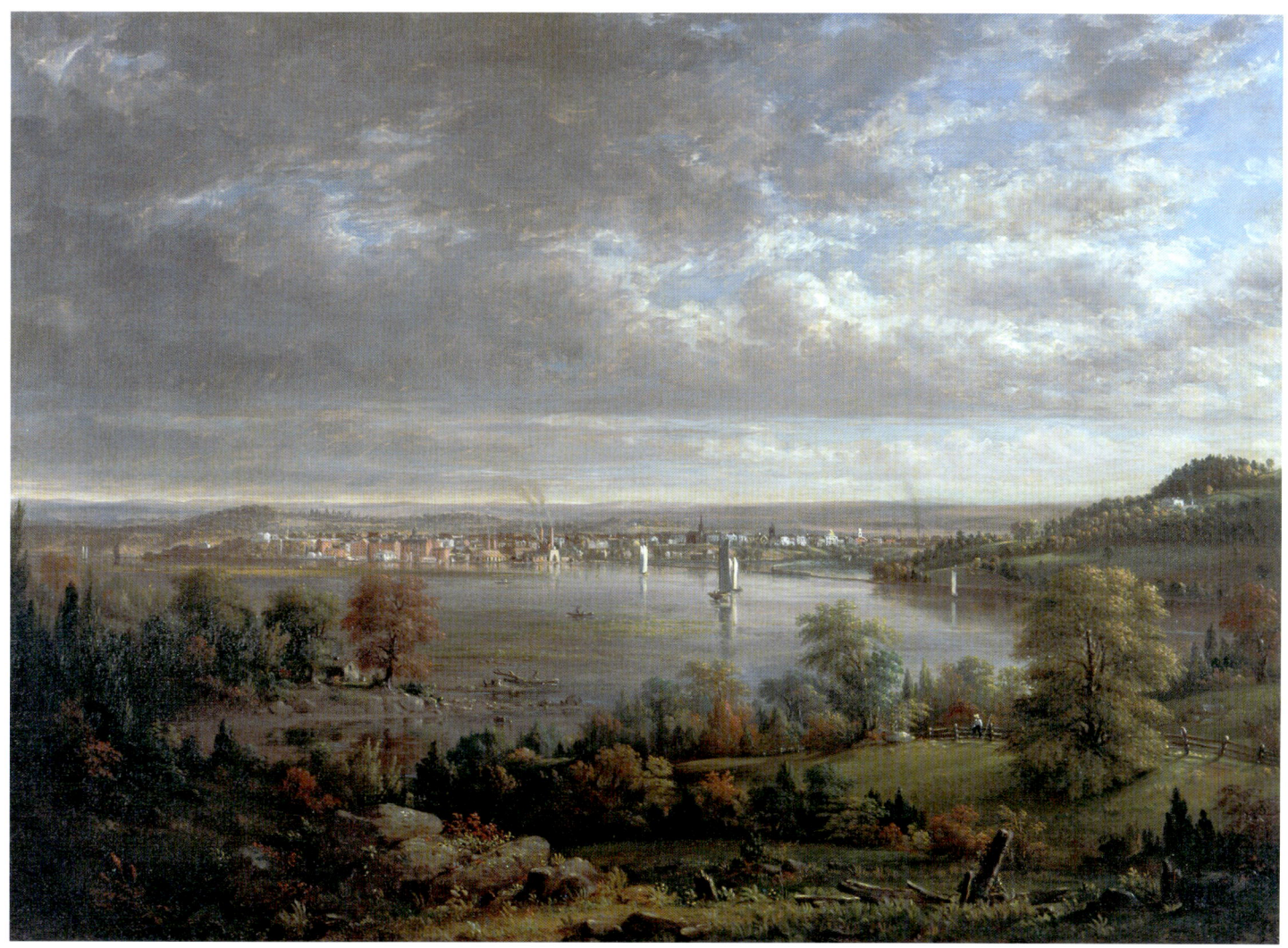

Fig. 35. Henry Ary (1807–1859). *View of the Hudson*, 1852. Oil on canvas, 26 × 36 in. (66.0 × 91.4 cm). Albany Institute of History & Art, Purchase (1977.17)

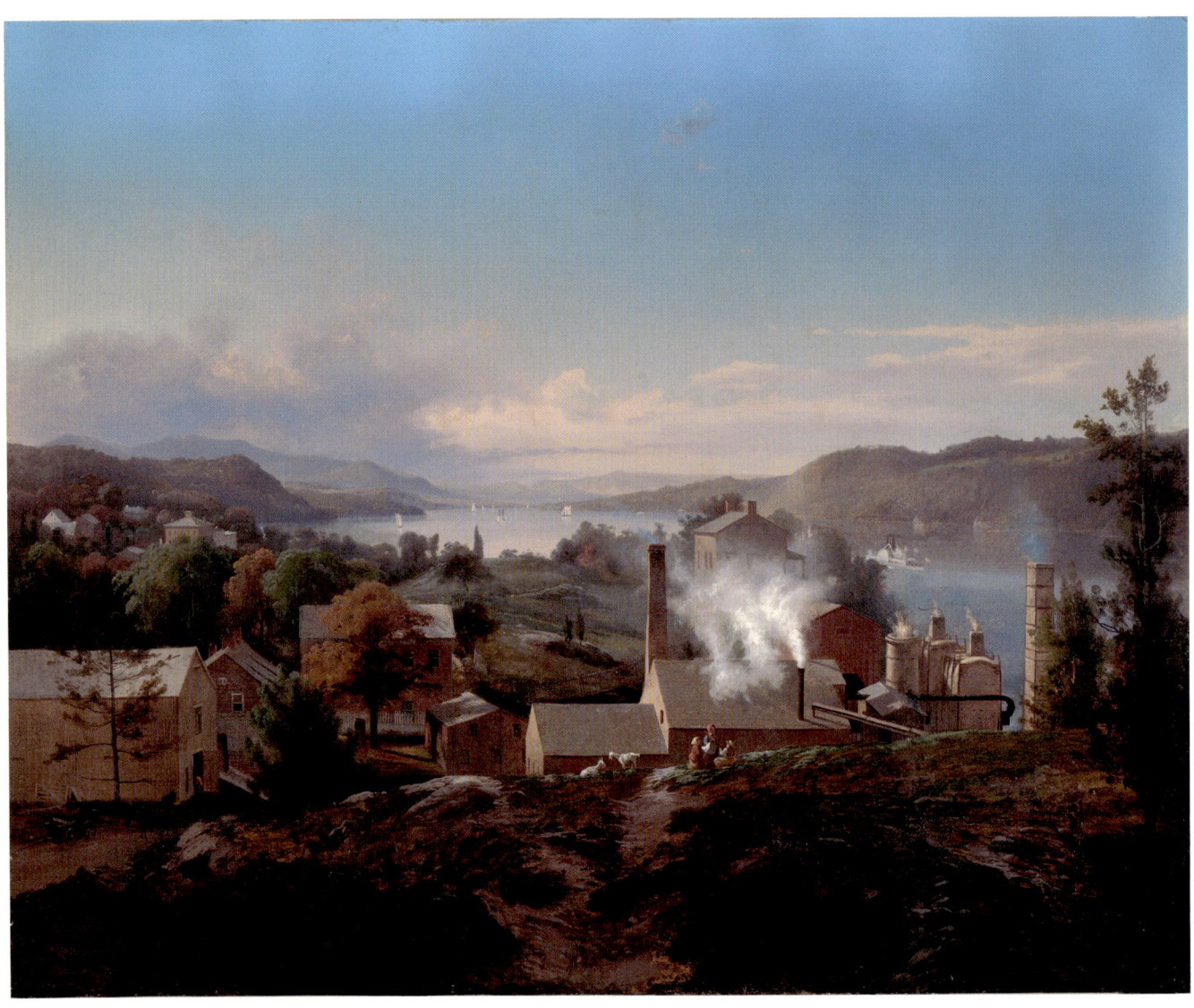

Fig. 36. Johann Hermann Carmiencke (1810–1864). *Poughkeepsie Iron Works (Bech's Furnace)*, 1856. Oil on canvas, 29 × 36¼ in. (73.7 × 92.1 cm). Yale University Art Gallery, New Haven, Bequest of Evelyn A. Cummins (1971.111.5)

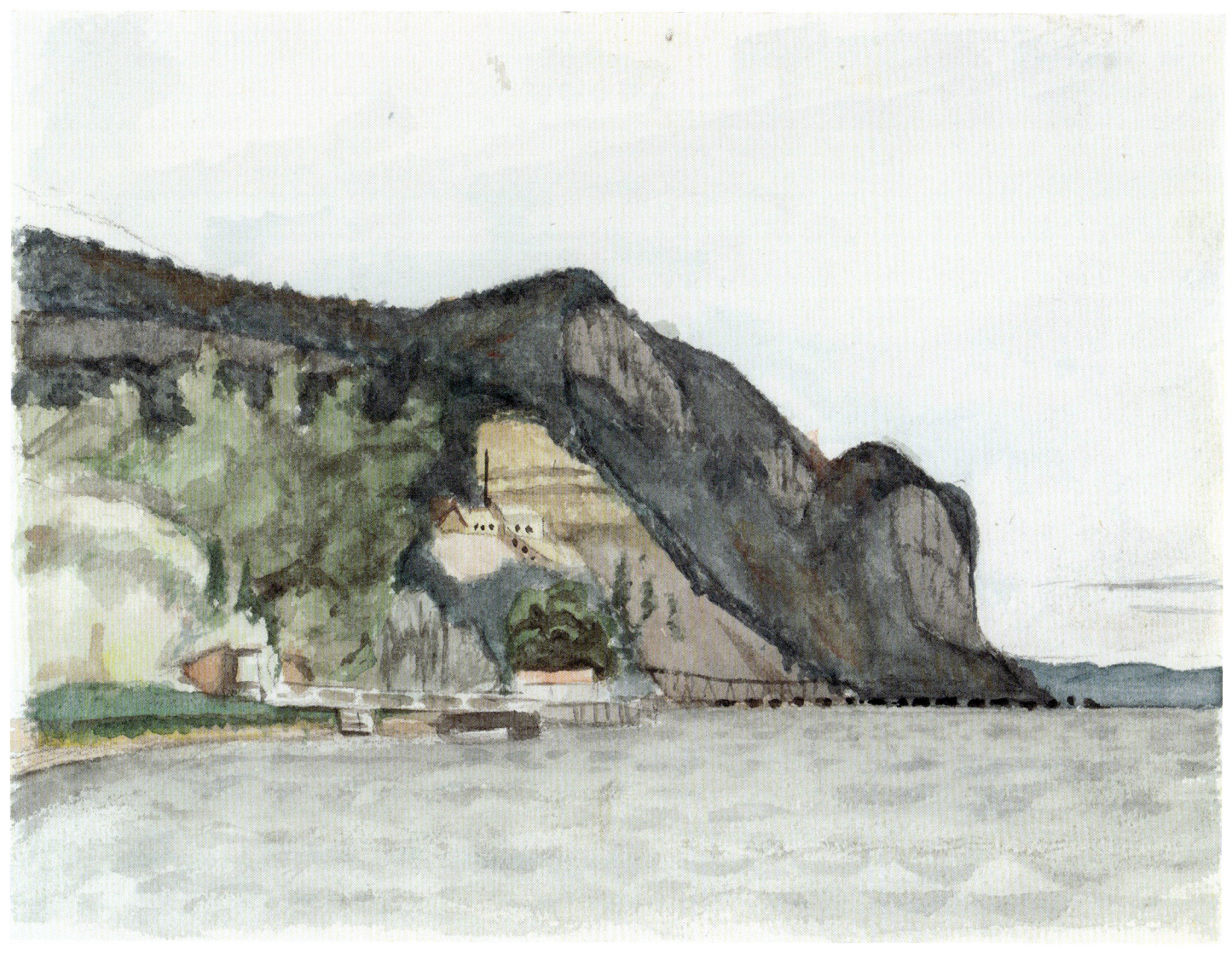

Fig. 37. Edward Hopper (1882–1967). *Hook Mountain, Nyack*, ca. 1899. Watercolor on paper, 5 1/16 × 6 7/8 in. (12.9 × 17.5 cm). Whitney Museum of American Art, New York, Josephine N. Hopper Bequest (70.1558.55)

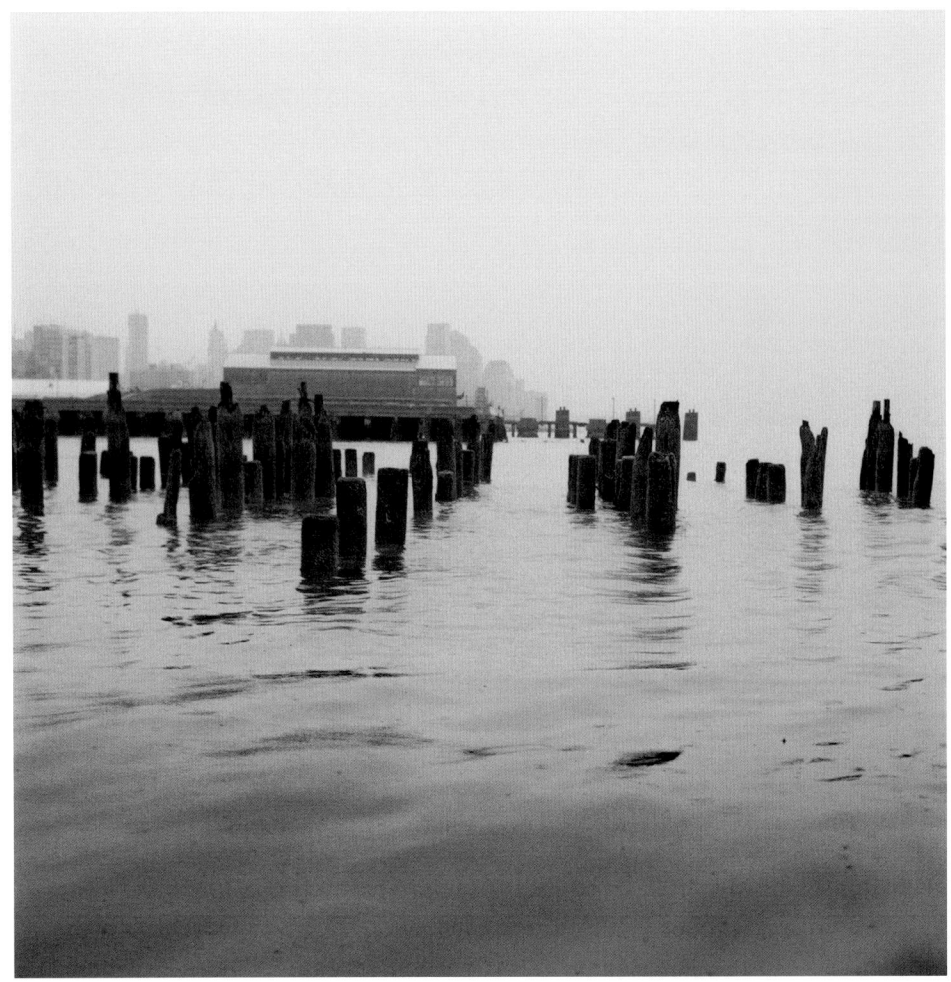

Fig 38. Every Ocean Hughes (b. 1977). *The Piers Untitled (#5)*, 2009–23. Gelatin silver print, custom frame, 33 × 33 in. (83.8 × 83.9 cm). Collection of the artist

Rosendale hydraulic cement industry, which produced more than half of the thirty-five million tons of natural cement in the United States.[22]

The air and water pollution continued unabated until after the middle of the twentieth century, visible in the rising heavy metals documented in the marshes. Robert Boyle describes the terrible state of the river, as untreated sewage from Utica traveled down the Mohawk to meet waste from factories, pulp wastes bobbing on the water, and decomposing sludge that acted like "a pistol pointed at the heart of the Hudson."[23] Turn-of-the-century quarries such as the one Edward Hopper portrayed in *Hook Mountain, Nyack* (fig. 37) were defacing the coastline, hardening the shoreline, and adding sediment to the river. In 1916 the Federal Dam at Troy was opened. This system of locks and dams artificially maintained a certain water depth. Dredging continued in the twentieth century, and the construction of dikes and the filling in of shores decreased the total water area by 30 percent when compared to the water area in 1910, as secondary channels decreased by 70 percent.[24] By 1932

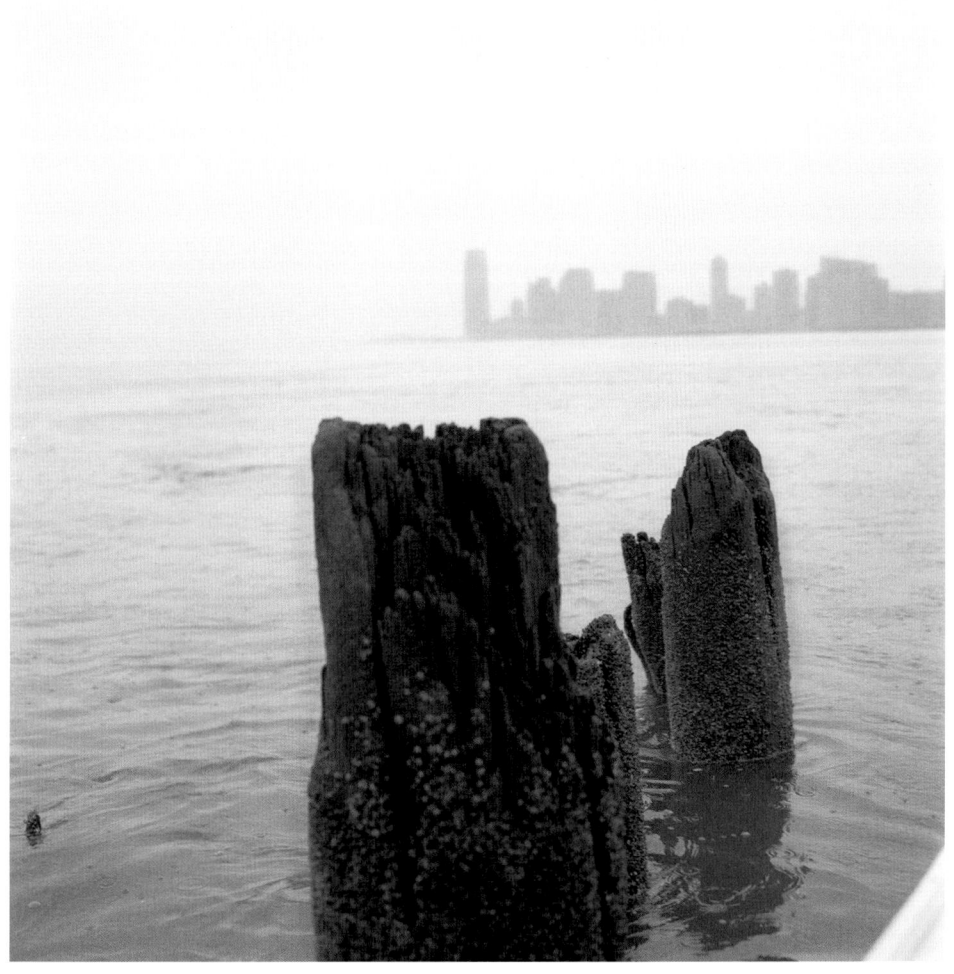

Fig. 39. Every Ocean Hughes (b. 1977). *The Piers Untitled (#7)*, 2009–23. Gelatin silver print, custom frame, 33 × 33 in. (83.8 × 83.9 cm). Collection of the artist

Congress approved dredging the river to a depth of about thirty-two feet, changing it from shallow and braided to deeper and channelized.[25]

This shoreline hardening is difficult to reconcile with the myriad inlets, channels, wetlands, and soft bays that once were present. Like the extreme and harsh monotonic concrete curves of Anthony Papa's *View of the Hudson River from Sing Sing 7 Building Yard* (fig. 45), the river's edge appears almost a wasteland. The loss of native biodiversity, including insects as well as humans, in Papa's *White Butterflies, Blue Hudson Painting* (fig. 44) is in stark contrast to the abundant life that rivers naturally represent.

While marsh records of the nineteenth and the twentieth centuries document increasingly heavy metal pollution of the Hudson, in the latter part of the twentieth century they also show zinc and lead declines.[26] In 1972 zinc pollution declined due to the passage of the Clean Water Act, followed in 1974 by reductions in lead due to the passage of the Clean Air Act with a ban on leaded gasoline. The eutrophication shows some improvement in the late twentieth century as well, with tertiary treatment starting in 1998 in New York City and further improvements in 2010. Yet, the waters of the Hudson retain high nitrates and are still absorbing heavy metals such as copper arsenate from the treated wooden piers that remain and have resisted corrosion for decades. Every Ocean Hughes's *The Piers Untitled #5* and *#7* (figs. 38, 39) depict these grim reminders of the watery concoctions that eventually make their way into the habitat that is so vital for the life of the estuary.

The Hudson marshes continue to sequester the heavy metals and lock up millennia of carbon-rich peat below the water composed of preserved plant matter known as "blue carbon" that is so important in preventing escape of decaying carbon dioxide rising to our atmosphere. These valuable ecosystems protect us from floods and dissipate wave energy, providing estuarine habitat and filtration for our water. As climate warms, sea level rises, and we gain access to our cleaner river margins, maintenance and expansion of these Hudson marshes will best sustain a healthy shoreline.

Notes
1. Steven H. Schimmrich, *Geology of the Hudson Valley: A Billion Years of History* (Amazon Digital Services, 2020); and D. M. Peteet et al., "Delayed Deglaciation or Extreme Arctic Conditions 21-16 cal. kyr at Southeastern Laurentide Ice Sheet Margin?" *Geophysical Research Letters* 39 (2012), https://pubs.giss.nasa.gov/abs/pe08400z.html.
2. R. E. Schmidt and T. R. Lake, "The Role of Tributaries in the Biology of Hudson River Fishes," in Jeffrey S. Levinton and John R. Waldman, *The Hudson River Estuary* (Cambridge: Cambridge University Press, 2006), ch. 15.
3. W. S. Newman et al., "Holocene Neotectonics and the Ramapo Fault Sea Level Anomaly: A Study of Varying Marine Transgression Rates in the Lower Hudson Estuary, New York and New Jersey," in D. Nummendal, O. H. Pilkey, and J. D. Howard, eds., *Sea Level Fluctuation and Coastal Evolution*, Society of Economic Paleontologists and Mineralogists Special Publication 4 (Bakersfield, Ca., 1987): 97–114; and D. M. Peteet et al., "Hudson River Paleoecology from Marshes: Environmental Change and Its Implications for Fisheries," in J. R. Waldman, K. E. Limburg, and D. Strayer, eds., *Hudson River Fishes and Their Environment*, A.F.S. Symposium 51 (Bethesda: American Fisheries Society, 2007): 112–28.
4. Alan W. Wells et al., "Temporal Changes in the Breeding Bird Community at Four Hudson River Tidal Marshes," *Journal of Coastal Research* (2008): 221–35.
5. Elizabeth Thompson and Dorothy Peteet, "Human Impact on Ramshorn-Livingston, a Hudson River Freshwater Tidal Marsh," in S. H. Fernald, D. J. Yozzo, and H. Andreyko, eds., *Final Reports of the Tibor T. Polgar Fellowship Program* (New York: Hudson River Foundation, 2018), sect. 7, 1–40.
6. D. E. Moerman, "Ethnobotany in Native North America," in Helaine Selin, ed., *Encyclopaedia of the History of Science, Technology, and Medicine in Non-Western Cultures* (Dordrecht: Springer, 2016).
7. Ibid.
8. Lucy Gill and Dorothy Peteet, "Reconstructing the Paleoecology of Haverstraw Tidal Marshlands," in D. J. Yozzo, S. H. Fernald, and H. Andreykyo, eds., *A Final Report of the Tibor T. Polgar Fellowship Program* (New York: Hudson River Foundation, 2015), sect. 5, 1–43.

9. S. Sritrairat et al., "A History of Vegetation, Sediment and Nutrient Dynamics at Tivoli North Bay, Hudson Estuary, New York," *Estuarian, Coastal, and Shelf Science* 102–3 (May 2012): 24–35, https://www.sciencedirect.com/science/article/abs/pii/S0272771412000741.
10. Milton M. Klein, ed., *The Empire State: A History of New York* (Ithaca and London: Cornell University Press, 2001); and Cynthia Owen Philip, *Rhinecliff: A Hudson River History, the Tangled Tale of Rhinebeck's Waterfront* (Hensonville, N.Y.: Black Dome Press Corp, 2009).
11. M. J. Collins and D. Miller, "Upper Hudson River Estuary (USA) Floodplain Change over the 20th Century," *River Research and Applications* 28, no. 8 (2011): 1246–53.
12. Daniel de Noyelles, *Within These Gates* (Thiels, N.Y.: self-published, 1982).
13. Daniel E. Miller, "Hudson River Estuary Habitat Restoration Plan," New York State Department of Environmental Conservation, Hudson River Estuary Program, 2013.
14. Theresa O'Meara, Suzanne P. Thompson, and Michael F. Piehler, "Effects of Shoreline Hardening on Nitrogen Processing in Estuarine Marshes of the U.S. Mid-Atlantic Coast," *Wetlands Ecological Manage* 23 (2015): 385–94.
15. Rachel K Gittman et al., "Engineering Away Our Natural Defenses: An Analysis of Shoreline Hardening in the US," *Frontiers in Ecology and the Environment* 13, no. 6 (2015): 301–7.
16. C. Frieswyk and J. Zedler, "Vegetation Change in Great Lakes Coastal Wetlands: Deviation from the Historical Cycle," *Journal of Great Lakes Research* 33, no. 2 (2007): 66–380; and D. Strayer and S. Findlay, "Ecological Functions of Hudson River Shorelines," *Aquatic Sciences* 72 (2013): 127–63.
17. S. M. Kercher and J. B. Zedler, "Flood Tolerance in Wetland Angiosperms: A Comparison of Invasive and Noninvasive Species," *Aquatic Biology* 80, no. 2 (2004): 89–102, doi: 10.1016/j.aquabot.2004.08.03
18. Cleo Chou and Dorothy Peteet, "Macrofossil Evidence for Middle to Late Holocene Vegetation Shifts at Iona Island Marsh, Hudson Valley, NY," in D. Yozzo, S. H. Fernald, and H. Andreyko, eds., *A Final Report of the Tibor T. Polgar Fellowship Program* (New York: Hudson River Foundation, 2015), sect. 2, 34.
19. D. C. Pederson et al., "Medieval Warming, Little Ice Age, and European Impact on the Environment during the Last Millennium in the Lower Hudson Valley, New York, USA," *Quaternary Research* 63 (2005): 238–49, https://pubs.giss.nasa.gov/abs/pe05200y.html; and D. Peteet et al., "Climate and Anthropogenic Controls on Blue Carbon Sequestration in Hudson River Tidal Marsh, Piermont, New York," *Environmental Research Letters* 15, no. 6 (2020), 065001, doi:10.1088/1748-9326/ab7a56.
20. K. Saltonstall, "Cryptic Invasion by a Non-native Genotype of the Common Reed, Phragmites australis, into North America," *Proceedings of the National Academy of Sciences* 99 (2002): 2445–49.
21. Pederson et al., "Medieval Warming."
22. D. Werner and K. C. Burmeister, "An Overview of the History and Economic Geology of the Natural Cement Industry at Rosendale, Ulster County, New York," *Journal of ASTM International* 4, no. 6, paper ID JAI100672.
23. Robert H. Boyle, *The Hudson River: A Natural and Unnatural History* (New York: Norton, 1979).
24. Collins and Miller, "Upper Hudson River Estuary."
25. Ibid.
26. D. M. Peteet et al., "Sediment Starvation Destroys New York City Marshes' Resistance to Sea Level Rise," *Proceedings of the National Academy of Sciences* 115, no. 41 (2018): 10281–6.

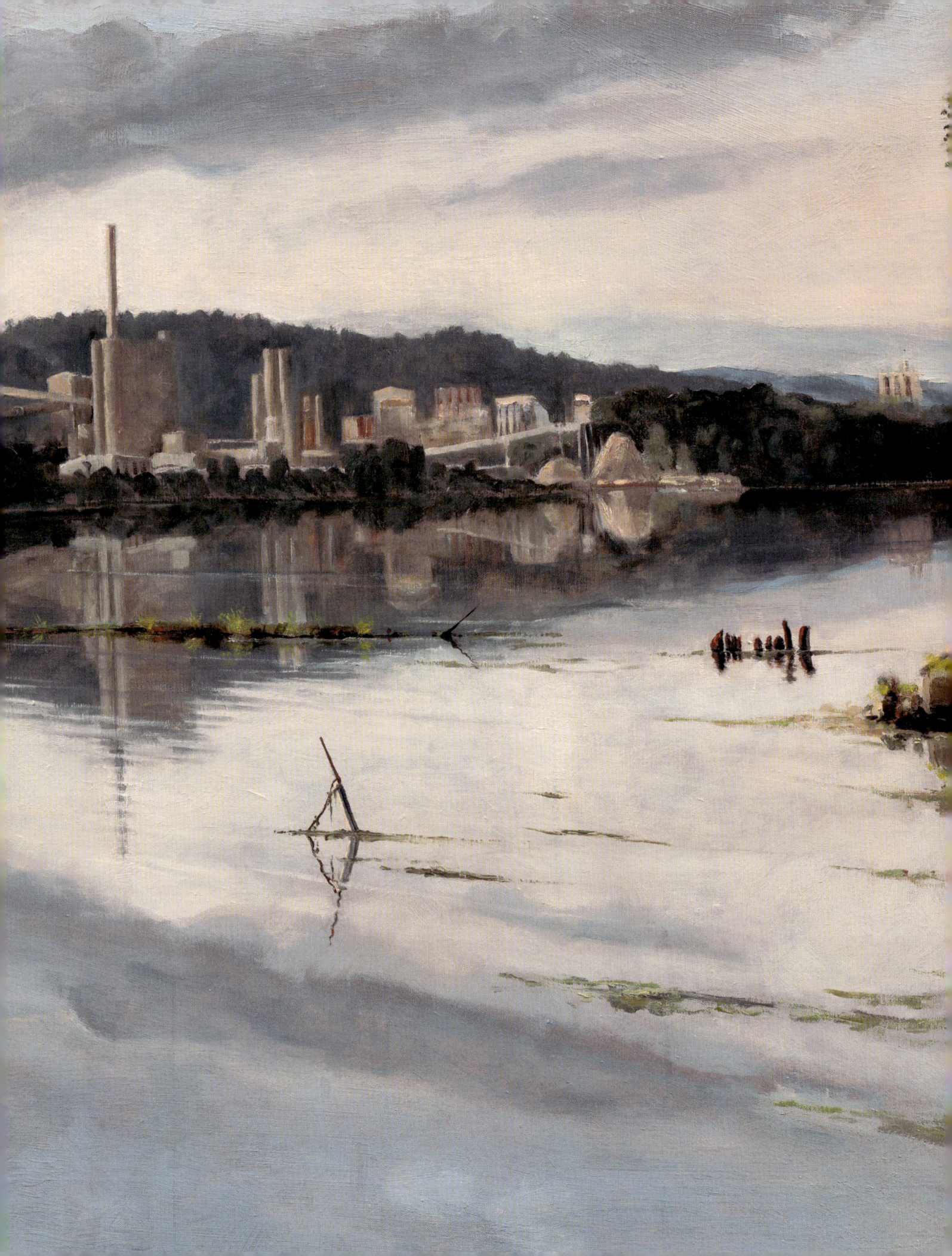

The Geology of Shifting Shorelines

Steven Schimmrich

There are two great themes in modern geology. The first is the concept of geologic time and the recognition that the earth has a 4.6-billion-year history. The second is plate tectonics, which is the understanding that the rigid outer part of the earth is split into several tectonic plates that move and fundamentally reorganize the planet over the immense span of geologic time.

Due to plate tectonics, the Hudson Valley has a complex geologic history involving the creation and eventual destruction of at least two supercontinents, multiple episodes of mountain-building as tectonic plates collided, and periodic inundations of seawater as erosion repeatedly brought the land surface back to sea level. During the past billion years, our area has moved from the Southern Hemisphere, crossed the equator, and drifted to its present-day location.

Throughout deep time, singular events have resulted in the formation of numerous geologic resources that have been integral to the history and development of the Hudson Valley. These resources include iron ore in the Hudson Highlands once vital to the Revolutionary War effort, metamorphic and igneous rocks formed deep in the roots of ancient mountains and now quarried as building stones, natural cement from Ulster County that literally built late nineteenth-century New York City, bluestone from the Catskills forming the curbs and sidewalks we walk upon, and bricks from the clay deposits of immense glacial lakes that once flooded the Hudson Valley to name only a few.

One such geologic resource is shown in Edward Hopper's watercolor of Hook Mountain (fig. 37). Overlooking the wide Tappan Zee section of the Hudson River, Hook Mountain was the location of a rock quarry in the Palisades Sill. Some two hundred million years ago, we were connected to Africa and part of a supercontinent named Pangaea. As Pangaea rifted apart, magma rose through the continental crust, spreading out between subsurface rock layers and cooling into a sheet of rock known as a sill. Much later this igneous rock, called diabase by geologists, was exposed along the west bank of the Hudson River as striking cliffs referred to by early settlers as the Palisades. Diabase is a hard, resistant rock once quarried for use as a durable base for roads and railroad beds. Quarrying on the face of the Palisades

Joellyn Duesberry. *Cement Factory, Hudson River* (detail), p. 79

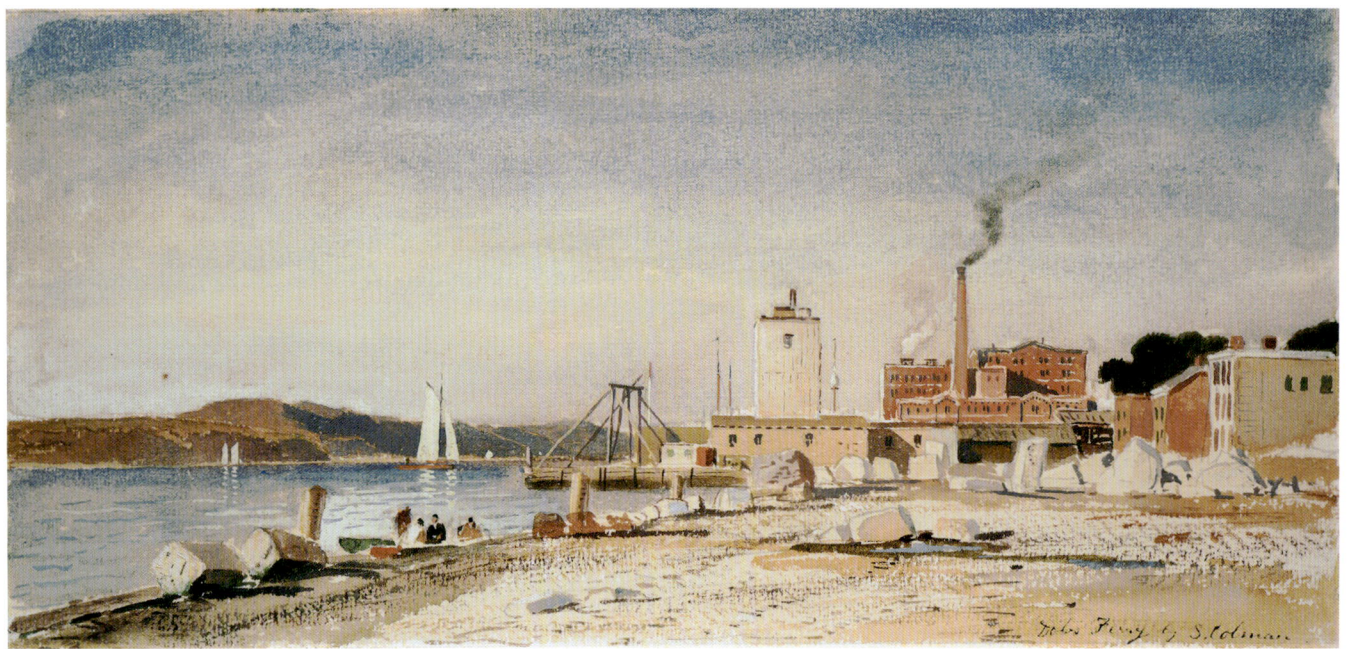

Fig. 40. Samuel Colman (1832–1920). *Quarry Works, Hastings on Hudson*, before December 1875. Watercolor on paper, 7 11/16 × 15 7/16 in. (19.5 × 39.2 cm). Museum of Fine Arts, Boston, Gift of Maxim Karolik for the M. and M. Karolik Collection of American Watercolors and Drawings, 1800–1875 (61.339)

mostly ended when those on the other side of the river realized it was marring their view.

In Samuel Colman's *Quarry Works, Hastings on Hudson* (fig. 40) we see quarrying of the metamorphic rock marble. This rock, variously referred to as Tuckahoe, Inwood, or Westchester marble, formed approximately 480 million years ago from sediments on a seafloor. Multitudes of marine invertebrates living on this seabed had hard parts made from calcium carbonate which, over time, decomposed into a carbonate mud. When this mud lithified into sedimentary rock, it became limestone. Groundwater later replaced some of the calcium with magnesium, creating a new rock called dolostone. Tens of millions of years later, a chain of volcanic islands collided to our east forming the Taconic Mountains, which once rivaled the Rocky Mountains in size. This collision heated and altered those buried dolostones into a dolomitic marble. This marble was later quarried as a building stone in various locations throughout Westchester County and used as façades on numerous structures, St. Patrick's Cathedral being a notable example.

In An-My Lê's *Trap Rock* series (fig. 62), we see the same rock unit but not quite metamorphosed into a marble due to an increased distance from that ancient mountain belt. Here, it is called the Wappinger Group carbonates and is quarried to be crushed into gravel. Gravel is more prosaic than its marble cousin farther south but forms the solid foundation of the roads and concrete structures we use in our daily lives.

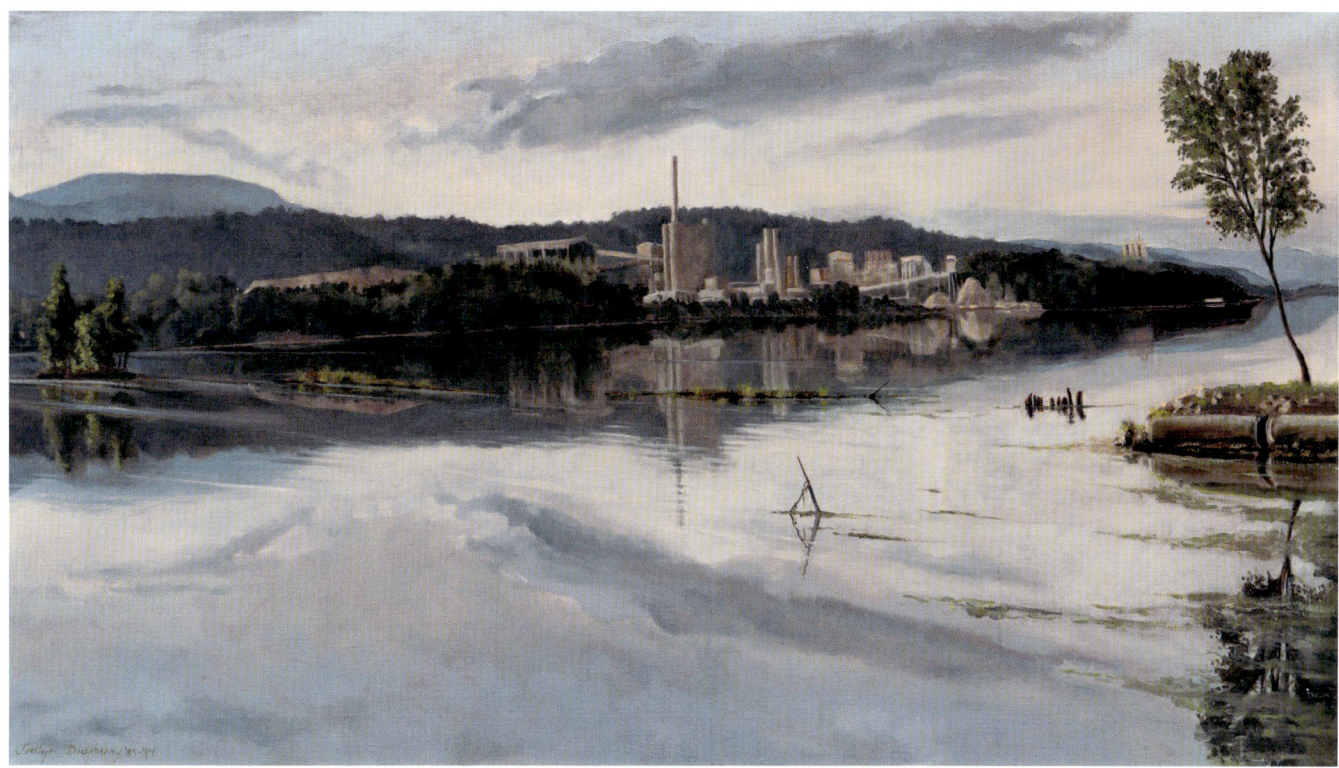

Cement Factory, Hudson River (fig. 41) by Joellyn Duesberry reveals an important geologic resource for the Hudson Valley. Cement begins with limestone or dolostone, the lithified sediments of an ancient seafloor, which is then heated in kilns and milled into powder. Some rock units, like the famous Rosendale dolostone that produced more than half the natural cement in the United States in the 1890s, had the exact chemical composition for making cement, while other limestone and dolostone layers can be made into cement only by adding certain missing elements during the manufacturing process.

While Hudson River School artists painted landscapes of a romanticized view of nature, the reality of the shifting shoreline was quite different. It was often a location of quarries and industrial sites unearthing geologic resources to feed the voracious growth and development of New York City as well as the towns strung along the banks of the mighty river.

Fig 41. Joellyn Duesberry (1944–2016). *Cement Factory, Hudson River*, 1983–84. Oil on linen, 20 × 36 in. (50.8 × 91.4 cm). Hudson River Museum, Yonkers, Gift of the artist, 2007 (2007.01.01)

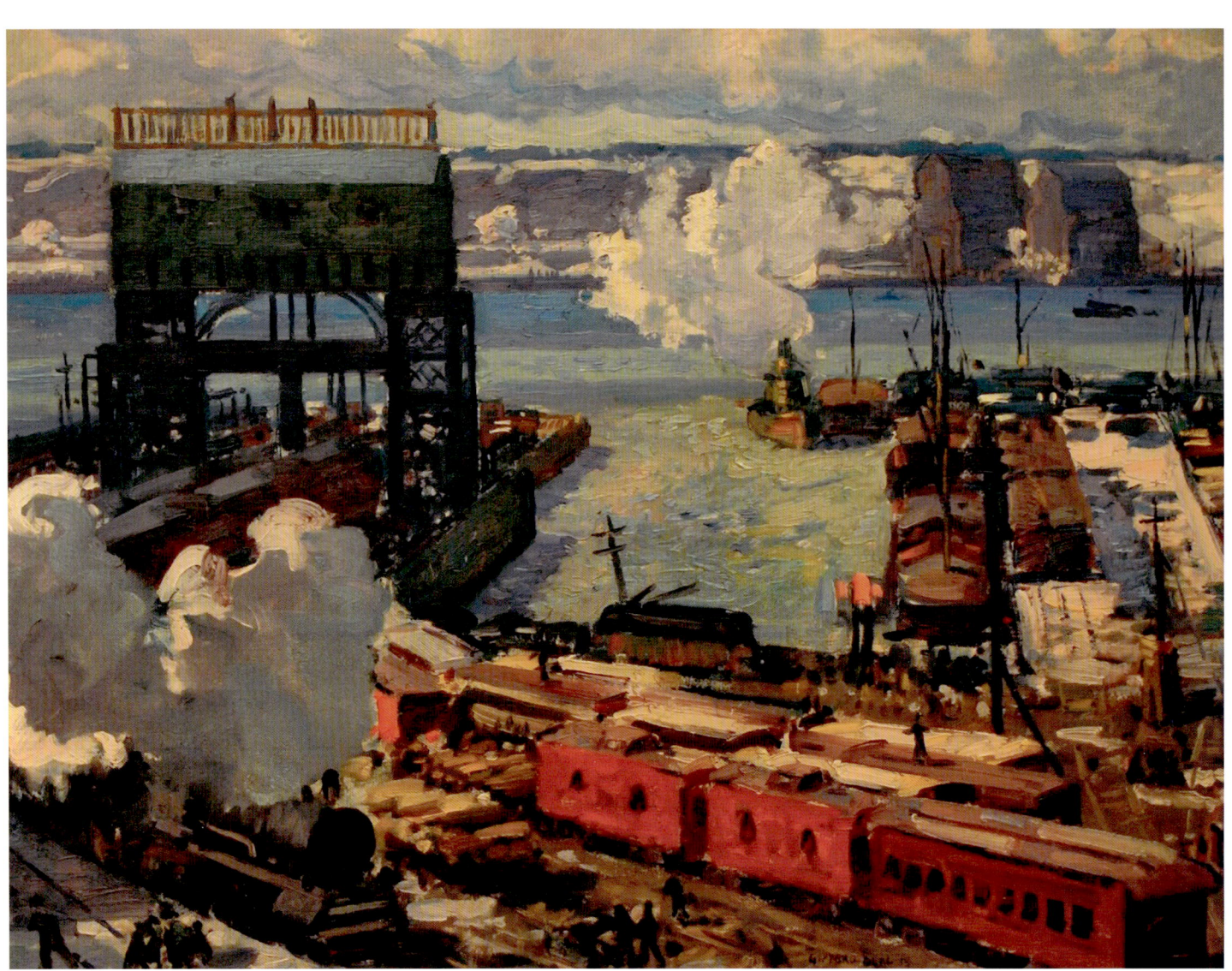

From Degradation to Restoration

Shifting the Hudson's Storyline

Tracy Brown

The reach of the estuary through the majestic Hudson Highlands offered a sea-level passage west, through the Appalachian mountain range, which became a vital commercial link between the Midwest and the East Coast. The riverbanks provided a level grade for railroad lines. The wide bays cooled power plants and carried waste from factories. The diverse species of fish provided profit until their numbers crashed from overharvesting, pollution, mill dams, and habitat destruction.

In landscape paintings, the river might look like the one silent, steady feature in a changing tableau of human activity along the edges—docks, factories, fishing boats, brickyards, quarries, freight yards, steamships, ironworks, landfills of ash and garbage. But the river has its own story of profound ecological changes driven by human activity.

Shifting Shorelines: Art, Industry, and Ecology along the Hudson River helps us see how human and river stories are intertwined. The works of art depict environmental threats that grew in severity during the centuries following European colonization and the Industrial Revolution. The mounting collateral damage—fish that tasted of oil, waters fouled by sewage, sediment saturated with toxic waste—drove a group of people in 1966 to organize the Hudson River Fishermen's Association, which later became Riverkeeper. They began, and we continue, to shift the storyline from degradation to restoration.

The steaming engines in George Bellows's *Rain on the River* (fig. 20) and Gifford Reynolds Beal's *Freight Yards* (fig. 42) foreshadow one of Riverkeeper's first cases: a lawsuit against Penn Central Railroad over discharges of oil from a rail-yard pipe into the Croton River. Samuel Colman's *Quarry Works, Hastings on Hudson* (fig. 40) and Jasper F. Cropsey's *The Hudson River at Hastings* (fig. 43) point to the location of another early case. Riverkeeper successfully took on Anaconda Wire and Cable, which operated at the Hastings waterfront from 1929 to 1975 using asbestos coating and mixtures of polychlorinated biphenyls, or PCBs, to make cables that were waterproof and fireproof. Thanks to a whistleblower who reported oil, metals, and solvents being dumped straight into the river, the company was prosecuted and fined. With legal representation by our partners, the Pace Environmental

Fig. 42. Gifford Reynolds Beal (1879–1956). *Freight Yards*, 1915. Oil on canvas, 36 × 48 in. (91.4 × 121.9 cm). Everson Museum of Art, Syracuse, Museum purchase with funds from American Art Fund (PC 15.80)

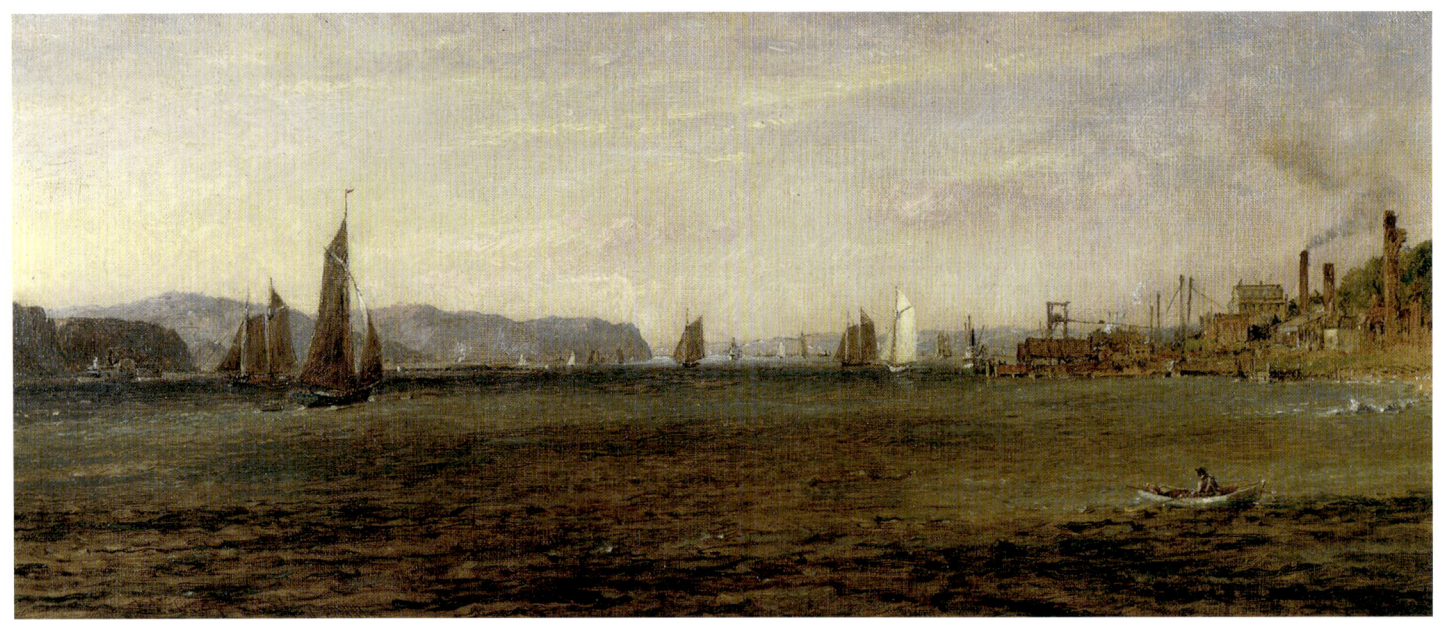

Fig. 43. Jasper F. Cropsey (1823–1900). *The Hudson River at Hastings*, 1889. Oil on canvas, 8⅝ × 20 in. (21.9 × 50.8 cm). Newington-Cropsey Foundation (168)

Litigation Clinic, Riverkeeper fought for decades for cleanup of PCBs and other waste at the site. The waterfront is finally ready for its next chapter.

Many of our oldest fights are ongoing. PCBs continue to poison fish throughout the Hudson due to dumping by General Electric between 1947 and 1977. Despite the dredging of highly contaminated sediment, PCBs remain a health hazard to those who eat the fish. Riverkeeper is actively fighting for additional cleanup measures.

Aaron Douglas's *Inwood Power Plant* (fig. 24) recalls another form of damage. Power plant intakes kill massive amounts of fish and eggs and discharge plumes of heated water. The Indian Point nuclear plant alone killed an estimated two billion fish and eggs every year it operated. With cleaner energy on the rise and the plant's shutdown in 2021—due in part to challenges by Riverkeeper to its permits and operation—life in the river has a better chance to rebound.

Despite pollution and a lack of public access, the Hudson maintains its irresistible draw, as seen in Ruth Orkin's *Boy Jumping into the Hudson River, NYC* (fig. 50). More than fifty years after the Clean Water Act, we are still working to reduce sewage pollution and fulfill the promise of swimmable waters.

Addressing pollution is not enough to restore the Hudson. We need to restore an abundance of life in the river. Scenes like Worthington Whittredge's *Shad Fishing on the Hudson* (fig. 16) call to mind the closure of commercial fisheries due to pervasive pollution and the continuing declines of our iconic species of fish. In the upper estuary, islands and shallows were destroyed long ago to improve navigation. We have started to reopen channels so that shad and other species find refuge and renew their populations. We are taking down old, obsolete dams so that Hudson Valley streams can thrive again with herring and eel. We are rebuilding oyster beds that clean the water, foster life, and build resilience.

Today's landscape includes Riverkeeper and our allies removing long-forgotten mill dams, hauling trash from the shorelines, monitoring water quality, and demanding more cleanup of toxic waste. The Hudson is full of opportunities for all of us to become good stewards and, generations from now, good ancestors, too. Let's paint that picture together.

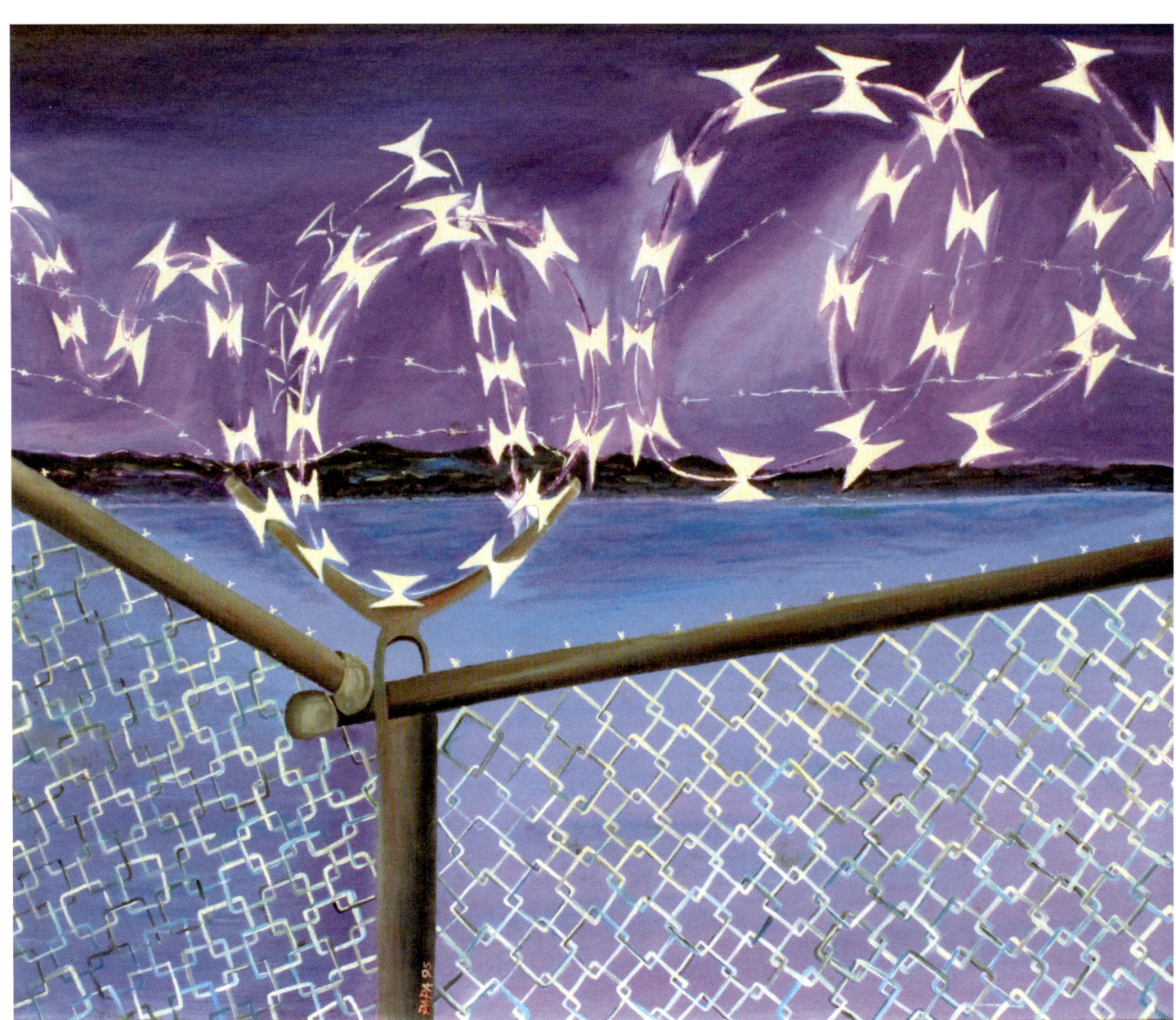

The Hudson River and Anthony Papa

Anthony Papa

I discovered my talent as an artist while serving a fifteen-to-life sentence for a nonviolent drug crime at Sing Sing Correctional Facility in Ossining, New York. In 1985, when I arrived at the prison, I was lost. Eventually, I found my talent through another artist who was imprisoned. My discovery of art helped me transcend the negative environment that held me captive. The canvas became an outlet for me to deal with my situation in a socially acceptable way.

I started reading and studying the lives of different artists. Because of my location, I was especially interested in the work of landscape artists such as Thomas Cole and Claude Monet. Sing Sing is surrounded by the Hudson River. My view of the river was seen through fences and thirty-foot-tall walls that were topped with miles of razor wire. The razor wire became a repetitive motif in my landscapes, akin to Monet's haystacks. The razor wire was drawn in many different sizes and shapes depending on how I felt that day. On a good day the razor wire was drawn small, becoming barely visible, as opposed to a bad day, when the prison environment overpowered me. On those days the razor wire was exaggerated in size and had a life of its own.

I found solace and peace in painting the Hudson River and even used the view as a calendar to mark my time. In the summer the sun bounced off the razor wire making it resemble the wings of butterflies in flight. It was in the yard of the prison's honor block, facing the river, that I captured on canvas the beauty of the Hudson River juxtaposed against the pathos of imprisonment (fig. 44). It was a double-edged sword seeing sail boats floating by in the water while I stood imprisoned on the grounds of the notorious prison. I thought it was part of my punishment for breaking the law, but in actuality my art became a survival tool that gave meaning and purpose to my life.

Being an artist in prison was not easy. Most art materials were considered contraband by the administration. No glass, metal, wood, or any flammable materials were allowed. Even painting views of the Hudson River was against the rules because they thought the images could be part of an escape plan. Despite the rules and regulations, I flourished as an artist and created a body of work that captured my environment.

Fig. 44. Anthony Papa (b. 1960). *White Butterflies, Blue Hudson*, 1995. Acrylic and oil on canvas board, 22 × 24 in. (55.9 × 61.0 cm). Collection of the artist

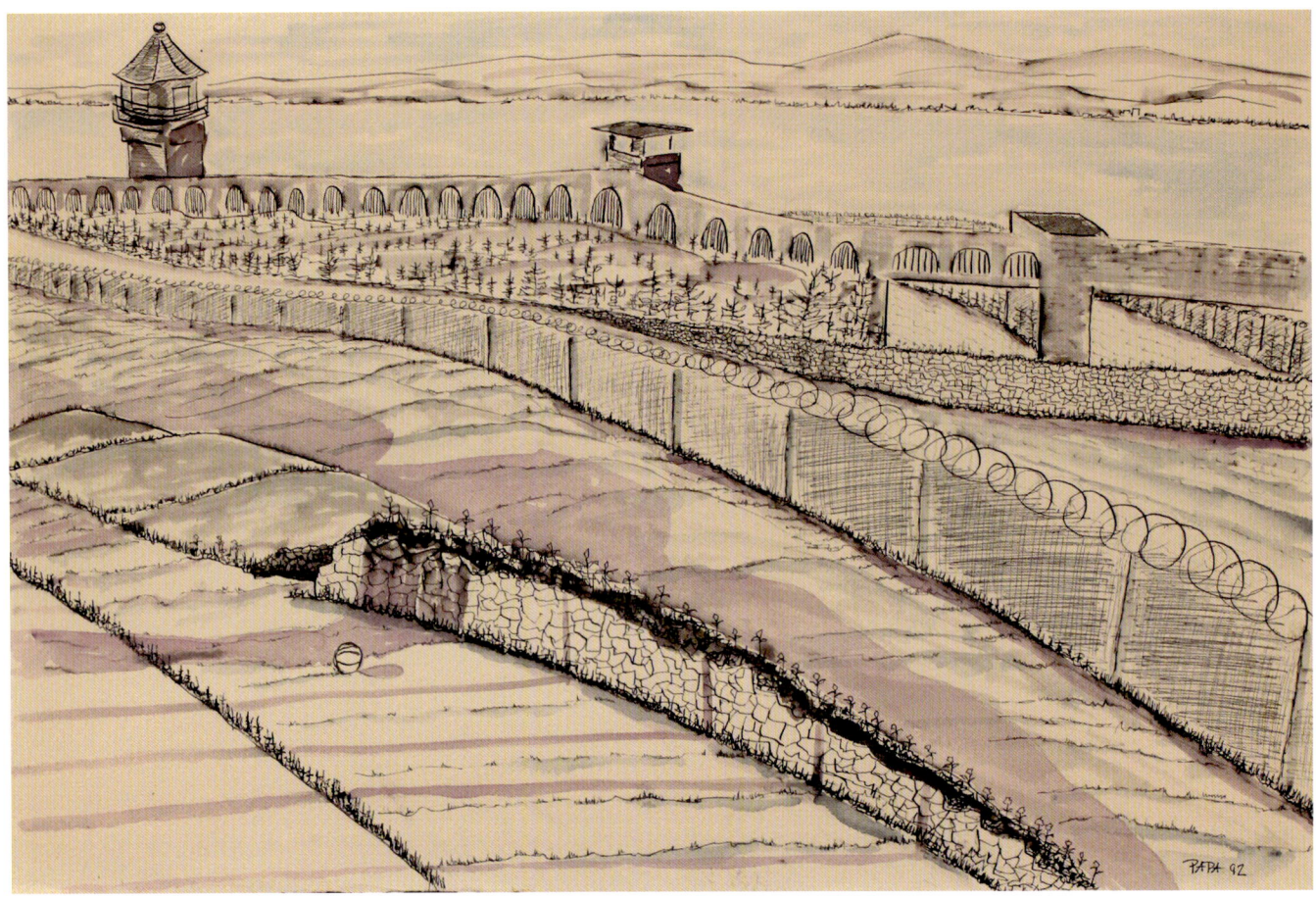

Fig. 45. Anthony Papa (b. 1960). *View of the Hudson River from Sing Sing 7 Building Yard*, 1992. Mixed media on 100% rag paper, 15 × 22 in. (38.1 × 55.9 cm). Collection of the artist

My art became very important in my quest to regain my freedom. In 1993 my life changed when the prison was contacted by Elisabeth Sussman, a curator at the Whitney Museum of American Art, asking to borrow a painting to be exhibited in the retrospective of conceptual artist Mike Kelley. My canvas *15 to Life* was chosen to be the centerpiece for Kelley's *Pay for Your Pleasure*.

In 1994 *New York Times* art critic Roberta Smith praised *15 to Life* as an "ode to art as a mystical, transgressive act that is both frightening and liberating, and thereby to intense and uncontrollable emotions of all kinds."[1] I received a tremendous amount of media attention, which led to me being granted executive clemency by Governor George Pataki. In 1996, after serving twelve years, I literally painted my way to freedom.

When released, I began exhibiting, using my art as a vehicle for social change. I painted about many social justice issues that affected society, ranging from the drug war to the environment.

One of the most rewarding experiences is when my art was used to help shut down Indian Point, a dangerous nuclear power plant near Sing Sing. In 2010 I was contacted by Hudson River Sloop Clearwater, an environmental group that wanted to use my prison landscapes as part of my testimony before the Nuclear Regulatory Commission. In the background of these landscapes, you can see two nuclear reactors jutting into the sky. It took a few years, but the plant eventually shut down operations.

Looking back at my landscapes of the Hudson River, I was amazed to discover that in Christian symbolism the razor wire that turned into butterflies represented the resurrected soul. Today, I continue to use my art as an agent of change and transformation, hoping to make the world a better place to live.

Note
1. Roberta Smith, "Mike Kelley's Messages: Mixed and Emotional," *New York Times*, November 5, 1993.

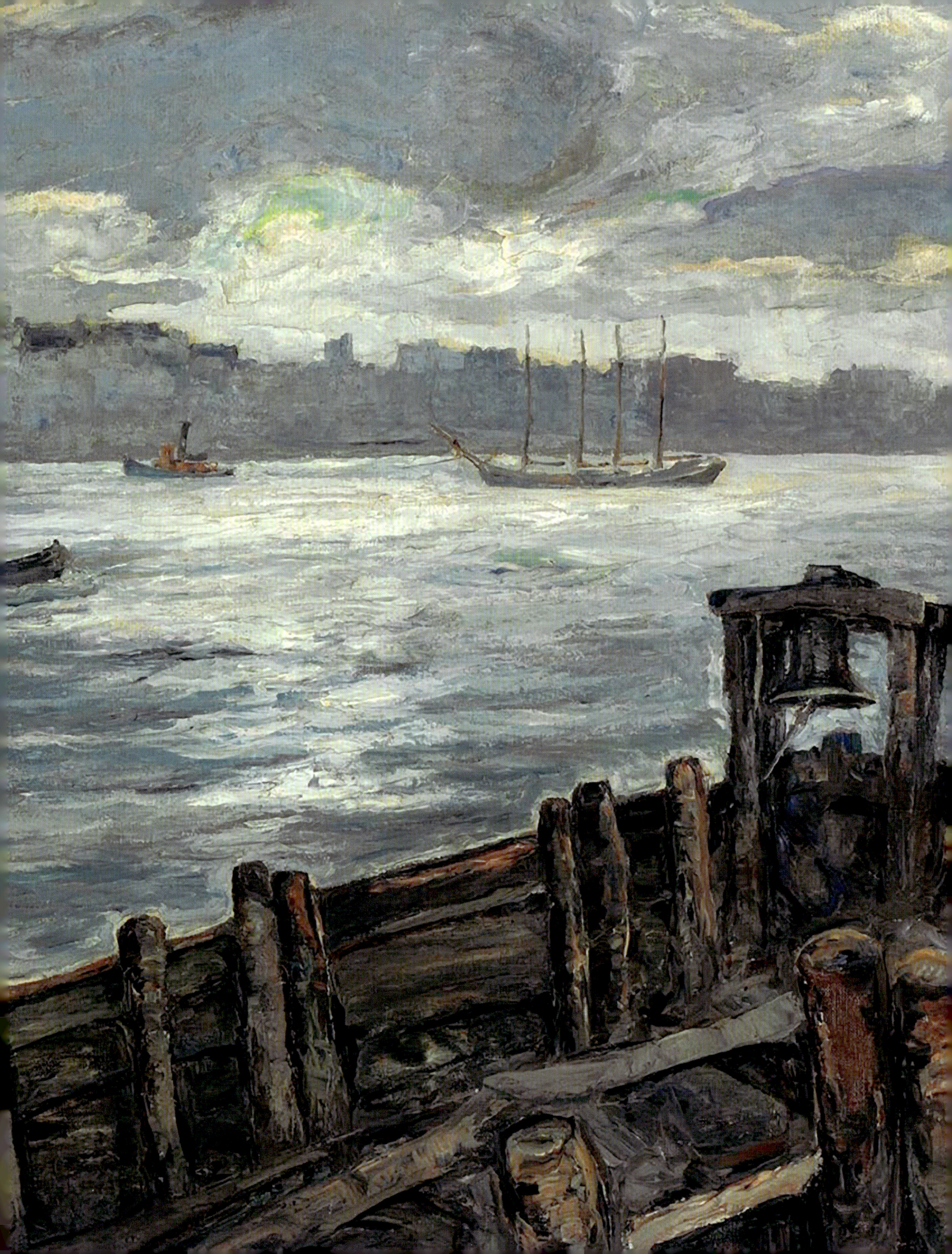

Hudson River as Sink

Waste, Pollution, and the Restoration of Nature

May Joseph

The Hudson is the first body of water in the history of environmental law around which questions of wilderness, pollution, and the idea of nature developed alongside a visual engagement with the river itself.[1] Inspired and sustained by the great river, Hudson River School painters depicted its ecosystem as an untouched wilderness that was fast disappearing in the wake of the thriving maritime and industrial economy of the nineteenth century. Interrupting this Edenic view of the Hudson River Valley as untouched before the encroachment of settlers and modernity, Robert Grumet writes, "Once, and not all that long ago, Manhattan was Indian country, as it had been for thousands of years."[2] Botanical[3] and archaeological[4] histories point to a longue durée of Native American disturbances shaping the wilderness. That settler colonialists approached the region as if it were a virgin forest is an indication of their imperiality toward Indigenous communities and nature.[5]

Shifting Shorelines: Art, Industry, and Ecology along the Hudson River presents an unflinching look at the river's anthropogenic pasts. It presents an assemblage of artistic, biotic, and scientific responses to the river's precarity. Materially grounded artworks engage a riverine aesthetics. They dialectically reframe the narratives of firstings and lastings,[6] such as the idea that non-Indians were the first to arrive on pristine vistas, that the Indigenous communities of the Hudson River Valley and New England had vanished. The multi-genre artworks disrupt the received assumption that the original inhabitants of the area had disappeared even as the descendants of different tribal nations continued to live in the Northeast.

Pollution is colonialism, writes Max Liboiron.[7] The colonial origins of pollution in the Hudson River and the Lenapehoking[8] (Lenape homeland) are documented by an exquisite gem in the exhibition, Kryn Frederycks's 1651 engraving *T'Fort Nieuw Amsterdam op de Manhatans* (fig. 3). The viewer is placed mid-river, floating above the archipelago, looking at the colonial town of New Amsterdam with the islands of Breuckelen[9] and Nuttin[10] in the distance. A flotilla of Lenni Lenape canoes rows upriver in the foreground as a Dutch barge approaches them downriver. At the upper right, three fluyts, the iconic three-masted Dutch naval ship of the seventeenth century, visualize

Palmer Hayden. *South Ferry* (detail), p. 92

their military power. Fort Amsterdam with its windmill and man-made shoreline occupies center stage.

Frederycks was the deputy and chief engineer of the New Netherland colony and designed the first Fort Amsterdam in 1626. That year Peter Minuit purchased Manahatta for $24 worth of commodities from the Munsee Indians. Frederycks's print is possibly the earliest documented representation of Dutch land use transforming the Munsee coastline. This rare record of New Amsterdam's shoreline hardening along its marshy ecosystem documents a scene of foundational contact, of colonial land grab. It signifies the ecological devastation of Lenape territory as the Dutch West India Company transformed the wooded cape of Lenapehoking on the shores of the Mahicannituck (Hudson) into the four-pronged Fort Amsterdam.[11] The image is a powerful testament to the colonial roots of pollution and extractive land use along the coast. It records the anthropogenic impact of colonialism on native ecology: a vast area cleared for military architecture by the colonizer. Manhattan's shores at the time were filled with marshes, woods, and brush. At night the sound of bears and other wildlife filled the air.[12] The construction of colonial forts across the former colonies has come to be understood as unaccounted for environmental degradation of native land use and ecosystems.[13] The scene freezes a critical moment in the narrative of modernity, that of the seventeenth century's European expansion onto Native American shores—in this case, the land of the Munsee.[14]

Hudson River visuality is a porous, muddy meshing of river and shore, salt marshes, tidal wetlands, creeks, and beach illustrated in Frederycks's depiction of an encounter between the Lenape and the Dutch on the Mauritse River, or the Noort Rivier, as the Mahicannituck was known.[15] Dutch hydrology techniques informed the early management of New Amsterdam wetlands.[16] Rooted in this history, the viewer is taken on an imaginary sail upriver and exposed to a painterly documentation of the factory pollution, industrial toxicity, and decaying piers of New York's polluted industrial waterfront for much of the last hundred years.

The ethical power of art in addressing the rising ecological concerns of waste and pollution around the increasingly degraded Hudson is addressed in the works of many artists by the end of the nineteenth century. The river is increasingly portrayed as both alluring and contaminated by extractive human activity. Questions of weather, density, toxicity, pollution, and social justice inform artistic investigations. The atmospheric 1898 lithograph *Hudson River, Oyster Market Near Christopher Street* by etcher and lithographer Charles Frederick William Mielatz (fig. 46) illustrates the dense commercial shipping activity around Christopher Street Pier, the largest pier in downtown New York. Mielatz's deft urban eye situates the viewer mid-river looking

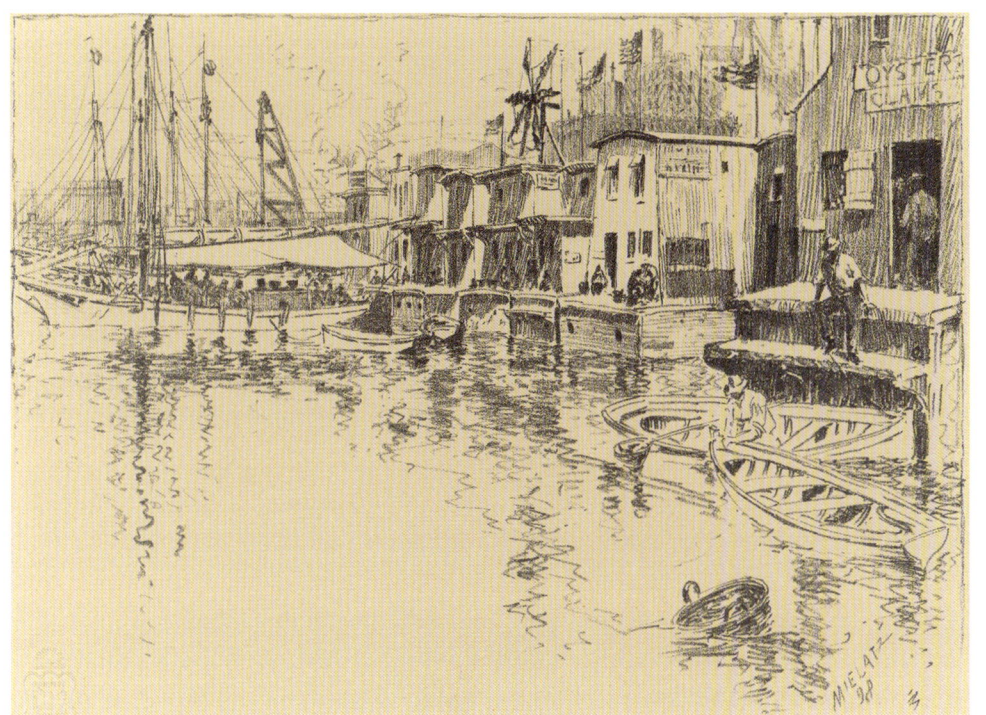

Fig. 46. Charles Frederick William Mielatz (1860–1919). *Hudson River, Oyster Market Near Christopher Street*, 1898. Lithograph, sheet: 7¾ × 10⅛ in. (19.7 × 25.7 cm). Museum of the City of New York, Arthur H. Scribner Collection, 1941 (41.62.20)

onto the shoreline. Renowned for his architectural etchings, Mielatz evokes a bustling culture of oyster fisheries, working longshoremen, a harbor of moored ships, and a submerged low-lying waterfront. Tenements housing dockworkers, sailors, riggers, and chandlers rise in the background. Mielatz's detailed observation of everyday events such as the floating basket of oysters in the foreground and his textured expressivity about river life presents an uncommon glimpse of late nineteenth-century river life. *Oyster Market* records the increasing toxicity of the Hudson River, leading to the closure of the last of its oyster beds in 1927.[17] Peter Malinowski, the founder of Billion Oyster Project, states that the New York harbor was a toxic environment for oysters until the 1970s and the passage of the Clean Water Act.[18] Mielatz's invaluable representation of the dwindling oyster industry offers a glimpse from the perspective of an oysterman of the untreated sewage and seeping pollutants that choked marine life.

John Marin's *Lower Manhattan from the River, No. 1* (fig. 27) portrays the same historic site of Frederycks's engraving, only three hundred years later. Marin was a friend of Alfred Stieglitz's and one of the most famous watercolorists of his time. His soaring metropolis is a towering, deforested coastal wall of skyscrapers and landfill. Marin grew up on the Jersey shore and lived in New York City.[19] His ebullient brushstrokes embody the rapacious mood of the New York shoreline. Devoid of nature as the Hudson harbor breakers crash against the hardened coast, *Lower Manhattan* portrays an ailing

HUDSON RIVER AS SINK

river choked by pollution. Raw sewage, paper mill and tannery discharges, and fecal coliform are the hidden contaminants that lead to an outbreak of typhoid fever in 1924. It was the era of accelerated aquatic degradation, with copper leaching from Brooklyn factories into the estuary.[20] Tempestuous, impressionistic, Marin painted a delirious toxic New York in all its stormy, bullish confidence before the crash of 1929.

Engaged with a contrasting view from the South Ferry Pier located ten minutes from Wall Street a year after the crash, *South Ferry* (fig. 47) by the prolific Harlem Renaissance seascape painter Palmer Hayden captures a gloomy, empty pier against the grays and cloudy blues of the Hudson River. The South Ferry docks are seaworn and in disrepair in the wake of the Great Depression. Shimmering lights of Brooklyn in the distance offer hope. Hayden is probably the most energetic riverine painter included in the show. One of the first Black artists to travel to Paris in the 1920s, he met Alain Locke, Countee Cullen, and other Black émigrés such as Hale Woodruff. Hayden struggled with navigating racism and class in the art world. Alongside a prodigious artistic life, he worked as a janitor in Greenwich Village during the 1930s.[21] His critique of the white art world and the invisible constraints on what Black artists could realize informs the contradictory trajectories of his work, bold and socially conscious yet sometimes deprecating.[22] While Hayden is best known for his stylized representations of Black urban

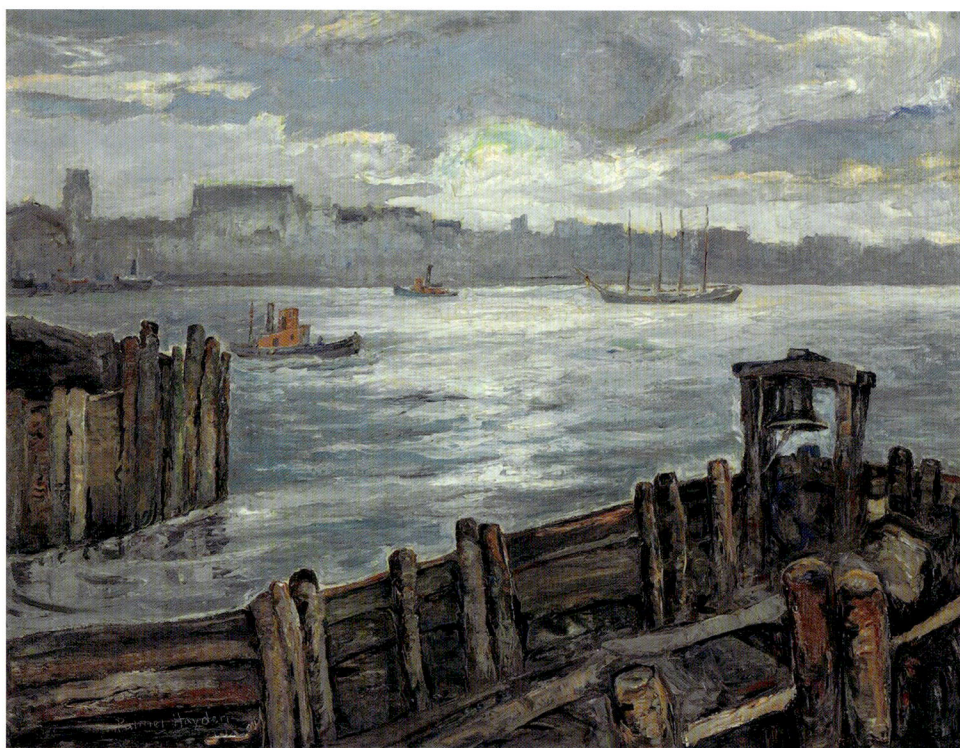

Fig. 47. Palmer Hayden (1890–1973). *South Ferry*, ca. 1930. Oil on canvas, 24 × 32 in. (61.0 × 81.3 cm). The Johnson Collection, Spartanburg, South Carolina

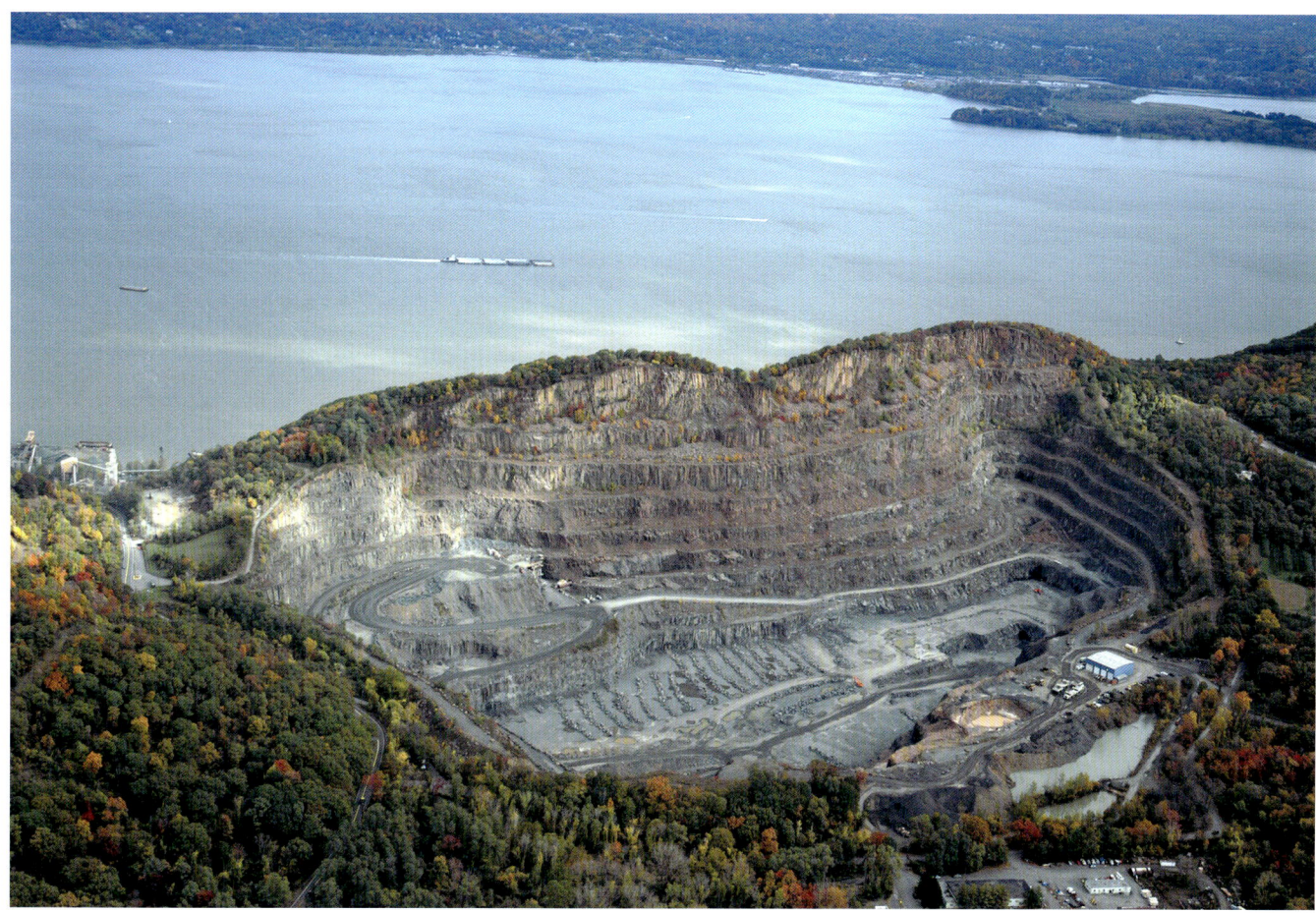

scenes, his vibrant, moody seascapes of the Hudson River add to the visualization of New York's environmental history. They reveal a loving chronicler of river pollution ahead of his time. Sulphurously prismatic, his detailed paintings of river life are the largest single visual repository of the heavy use to which the Hudson River was put in the early twentieth century. Shipping is one of the primary sources of ocean pollution. Blackwater, gray water, air, oil, and acoustic pollution are all part of the intensive degradation of riverine and coastal communities weighed down by shipping ballast water discharge alongside sulphur, diesel fuel, and eutrophication.

Attentive to the deteriorating hydrology of the Hudson River estuary is the searing sequence of photographs by the Center for Land Use Interpretation, *Up River: Man-Made Sites of Interest on the Hudson from the Battery to Troy* (fig. 48). A photojournalistic journey, *Up River* provides a visual vernacular for the Hudson's industrial and infrastructural history. Aerial and mid-river shots of chemical and cement factories, quarries, automotive junkyards, and hydroelectric plants record the transformation of the majestic shoreline

Fig. 48. Center for Land Use Interpretation. "Haverstraw Limestone Quarry," from *Up River: Man-Made Sites of Interest on the Hudson from the Battery to Troy*, 2006
Digital photograph
The Center for Land Use Interpretation, Culver City, California

along both the New York and New Jersey coasts. *Up River* showcases the most egregious sources of the river's effluence. It offers context for how the Hudson River Valley transformed from the wilderness captured in the works of nineteenth-century romantic landscape painters to the heavily contaminated sacrifice zones of dying marine life in the mid-twentieth century.[23] *Up River* juxtaposes the mythical stature of the Hudson in the American imagination with its transformation into an ecological sink for waste and contamination from Troy all the way to the largest metropolis in the world by the 1960s, New York City.[24] The bird's-eye views and aerial photographs shock the viewer into understanding the visionary melancholia of the Hudson River School's preoccupation with a disappearing wild. The grim yet scenic photographic sequence is a backdrop to the hard-edged optimism of subsequent artists responding to a degraded marine and archipelagic ecosystem in the Hudson/Raritan estuary.

CLEAN WATER, CLEAN AIR, AND THE RESTORATION OF NATURE
As the Hudson was the first contaminated river to be gradually cleaned and repaired thanks to the Clean Water Act of 1972,[25] its significance in the history of water laws is far reaching. The environmental struggles to restore habitats and shift industrial pollution toward ecological restoration resulted in new laws around the management of clean air and water. Unfortunately, sites that have benefited from this law are now threatened by the 2023 Sackett decision that reversed federal protections for countless wetlands in the United States.[26] Several artworks examine the environmental efforts transforming the Hudson River such as constructed landscapes along its shorelines, reviving marshes, dunes and beach replenishment, and rewilding.

In a series of unforgettable photographs of iconic shorelines, the queer performance and visual artist Every Ocean Hughes offers a conservationist's view in *The Piers Untitled* series (figs. 38, 39). Hughes's eye is experimental, dystopic, and lyrical. Complementing Alvin Baltrop's photographic series *The Piers* (figs. 54, 55), Hughes presents a climate-attenuated, submerged view of the city's disappeared queer locales along the Hudson that were especially popular in the 1970s and 1980s. Her fragmented and tilted perspectives of the distinctive wood pilings that are now active sites of aquatic rewilding chronicle loss and the AIDS crisis. Hughes's immersive camerawork is embedded in the Hudson River. Wide-angled frames of the queer community's disappearing archives are memorialized in monumental friezes against the New York skyline. Moody and river bound, the partially overexposed photographs present an off-kilter shore. The river is now a place of respite, allowing a photographer into its much cleaner waters. Hughes's

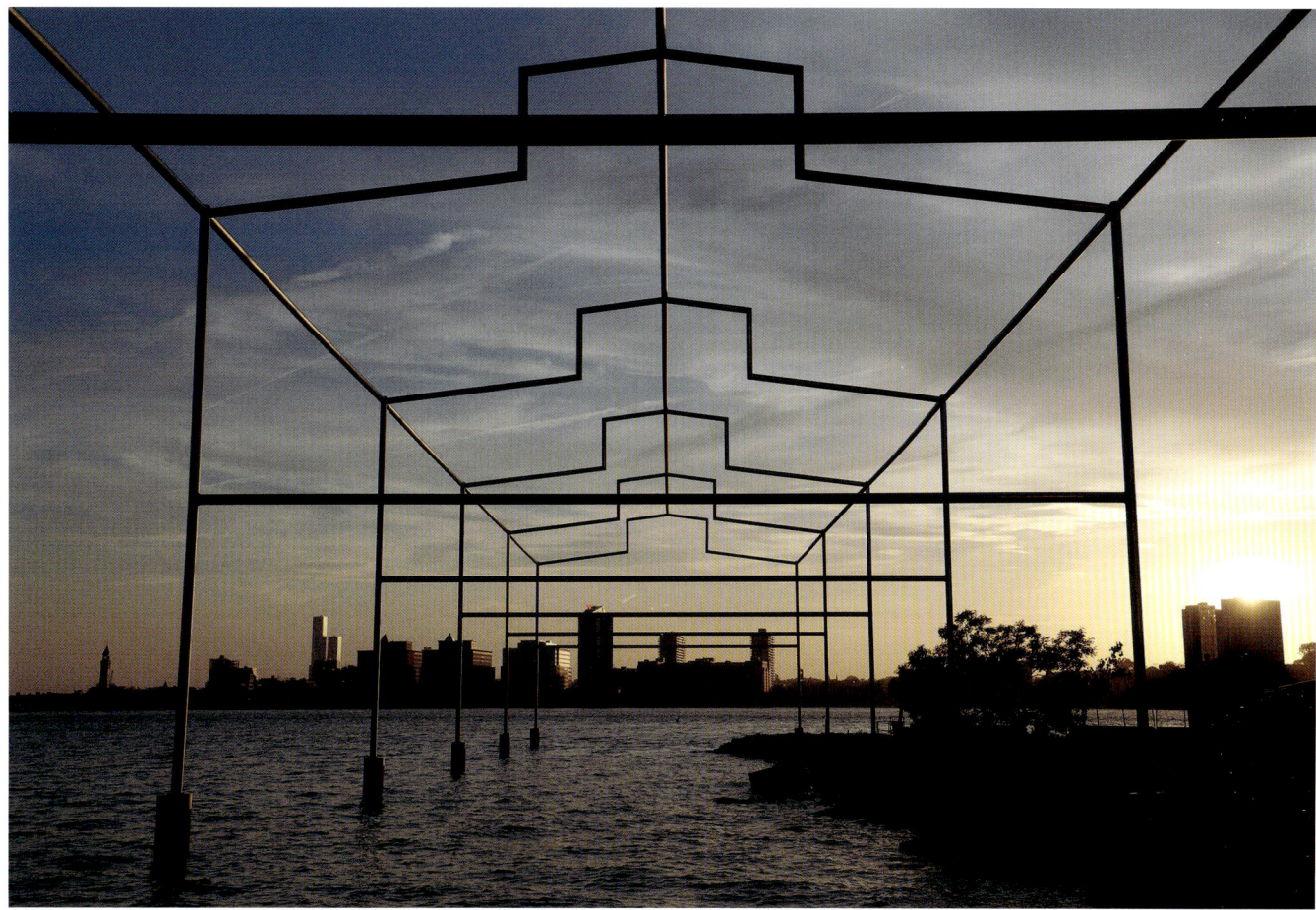

Fig. 49. David Hammons (b. 1943). *Day's End*, 2014–21. Siite-specific installation, Hudson River Park. Photo by Jason Schmidt for Whitney Museum of American Art, New York

images, in contrast, are epitaphs to loss, death, and remembrance of queer New York's history. Her lyrical documentation of what is gone also marks an emerging, resilient Hudson River that is gradually being repaired.

David Hammons's *Days End* (fig. 49) marks the palimpsestic histories of the dilapidated wooden piers at Gansevoort Peninsula in Manhattan's West Village. The controversial site-specific installation is a submerged metallic temple to water and light. The columns reflect the ocher afternoon light reminiscent of an Asher Durand painting as ring-billed gulls frolic between the steel pilings. Sponsored by the Whitney Museum of Art and located at the corner of 14th Street and the Hudson River, Hammons's art installation is a play on Gordon Matta Clark's 1975 cutouts into the abandoned built environment of Pier 52. Where Matta Clark's works commented on the urgency of pollution, contamination, and ephemerality of shoreline architecture, Hammons's *Day's End* is an elegy to the Hudson's restorative history in its post-industrial phase. The installation points to multiple ecological concerns raised by any construction being inserted into the river's ecosystem. Its minimal design avoids further

HUDSON RIVER AS SINK

marine disruption even as it invokes the industrial past of downtown New York's now-defunct port infrastructure alongside Pier 52's queer memories.[27]

In 1609 the river's edge was Greenwich Street and the site of the largest Lenape settlement of Sappockanican.[28] Eric Sanderson pointed out that industrial landfill extended the island outward toward now-defunct 13th Avenue,[29] creating the new fake landscape of the Gansevoort Peninsula whose history of garbage and toxic dumping into the river contributed to the establishment of the Clean Water Act.[30] The submerged infrastructure of *Day's End* is a reinsertion of that critical riparian history. It is a provocation to think through what Timothy Morton calls the ecological mesh, the awareness that all beings are interconnected.[31]

In the 1970s Matta Clark drew attention to the mesh of pollution that surrounded the noxious incineration Gansevoort Peninsula Destructor Plant that dumped its contaminants and toxic waste into the Hudson River. David Hammons's 2021 installation redraws attention to these histories of waste and pollution that led to the Clean Water Act and the Clean Air Act and the subsequent dismantling of the Gansevoort incinerator. Hammons's sculpture reopens a dialogue between the dramatic transformation of the Gansevoort landfill from the elevated Westside Highway to the era of rapidly transforming ecological parks. Experiments in afforestation and climate adaptation are emerging along the shoreline as climate mitigation initiatives are generating fake landscapes of sandy beaches, hills, dunes, oyster beds, and rewilding refuges. The Little Island, an artificial park mushrooming out of the Hudson River is an added enigmatic presence on the shores of the nineteenth-century landscape created out of landfill.

REWILDING THE RIVER

The Hudson River has been on the forefront of marine rewilding efforts in the United States since the Clean Water Act. Along its landfilled waterfront are growing examples of low-impact resiliency plantings and the return of swamps and marshlands in a climate-sensitive approach to restoring nature in the estuary. New trees and green spaces along the river's shores are part of the climate adaptation techniques being incorporated into the riverine biosphere. This resiliency can be traced to a singular event, 9/11, that forced a rethinking of New York City's relationship to its river culture.

Ruth Orkin's startling photograph *Boy Jumping into Hudson River, NYC* (fig. 50), set on a West Side pier, brings a haunting déjà vu of 9/11 to those who lived through the collapse of the World Trade Center. Orkin's improvisational low-angle shot and feel for the pier is breathtaking. Broken windows frame playful youth sunbathing at an abandoned industrial site. Jumping into the river—the water was clean enough—evokes an optimism that we

Fig. 50. Ruth Orkin (1921–1985). *Boy Jumping into Hudson River, NYC*, 1948. Gelatin silver print, 14 × 10 15/16 in. (35.6 × 27.8 cm). The Jewish Museum, New York, Purchase: Horace W. Goldsmith Foundation Fund (2008.98)

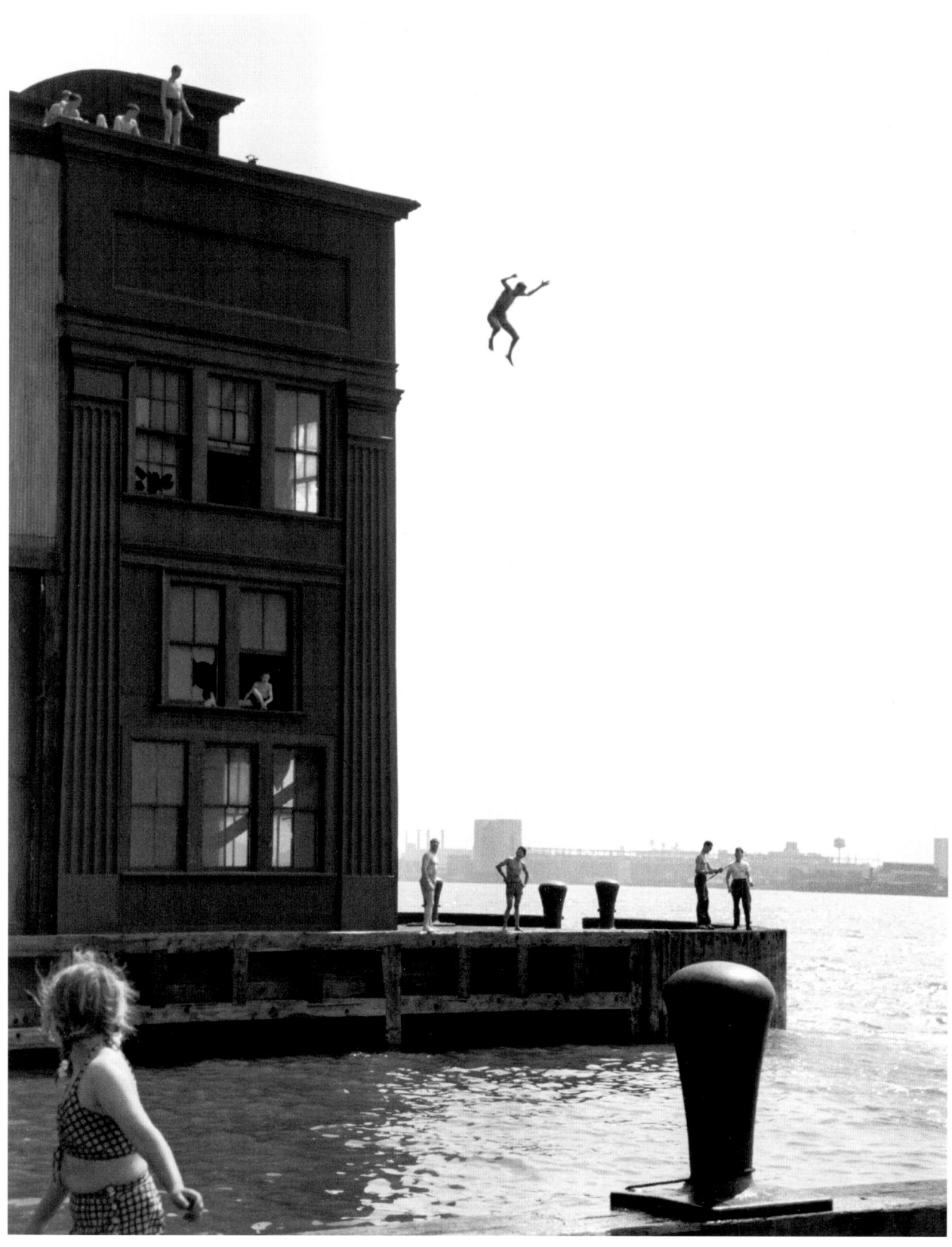

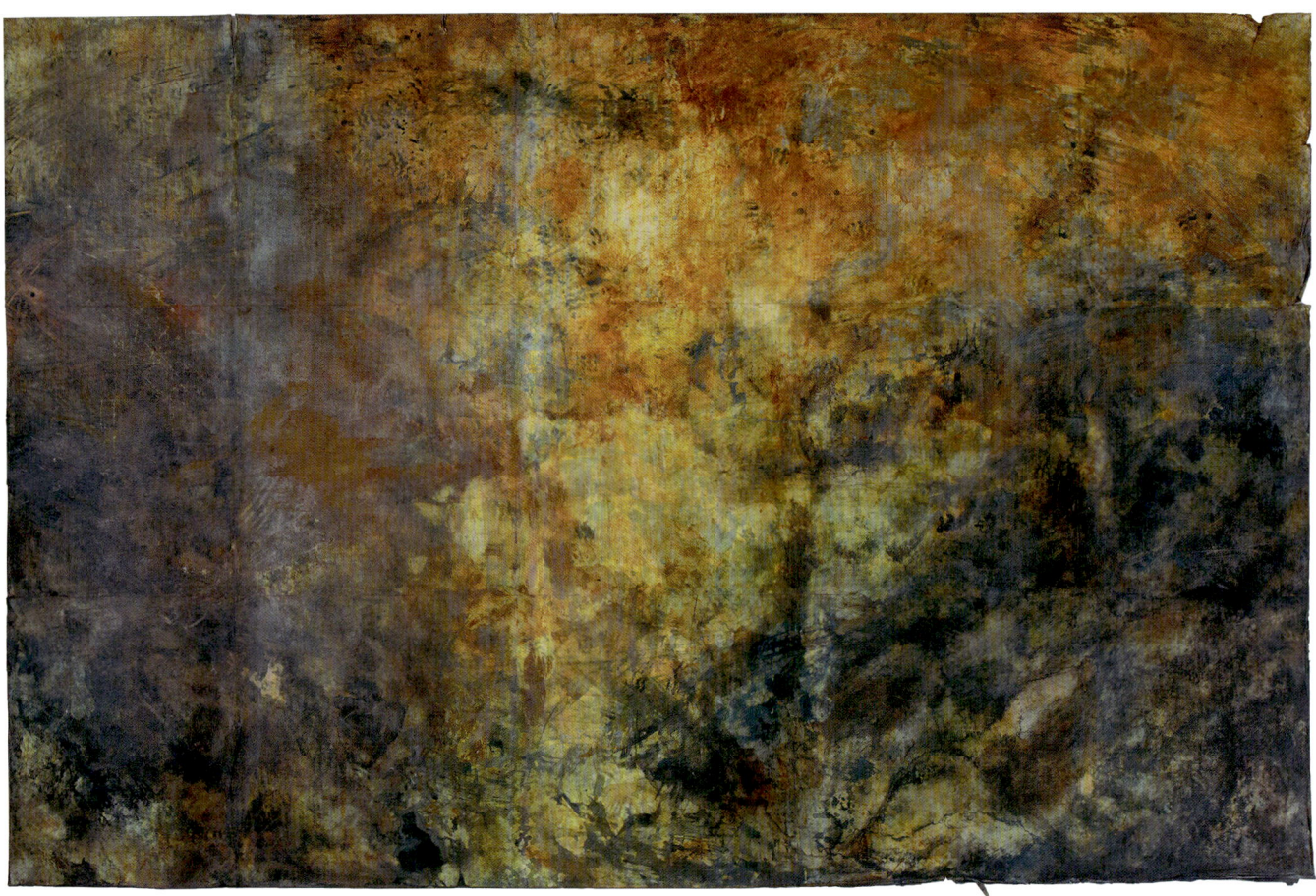

Fig. 51. Athena LaTocha (b. 1969). *Burning, Sulphuric, Violent*, 2020. Shellac ink, World Trade Center building sand, on paper, 132 × 204 in. (335.3 × 518.2 cm). Virginia Museum of Fine Arts, Arthur and Margaret Glasgow Endowment

cannot share in 2024 because the Sackett decision relegates waterways and wetlands much more vulnerable to the negative impacts of abuse by development and other polluting interventions. The fight to protect the Hudson's headwaters and small streams reminds us that the river's slow healing is always fragile and vulnerable to political winds.

Burning, Sulphuric, Violent (fig. 51) by Alaska-born Indigenous artist Athena LaTocha is a visceral citation of 9/11. Her technical virtuosity wielding metallic and industrial waste drawn from the 9/11 site into a living painting imbibes what Robin Wall Kimmerer calls "a grammar of animacy."[32] *Burning* is a magisterial embossing of World Trade Center rubble in an exploration of contamination on human and nonhuman livability. LaTocha's immersion in demolition sites, industrial ruin, and toxic debris brings a raw edge to the literal remnants of Hudson River landfill used to create a landscape painting. In *Burning,* inanimate nature, the landfill detritus of the polluted site, is recycled into a material abstraction of paint and light.

Art theorist Lisa E. Bloom observes, "One of the most important challenges facing artists, activists, and scholars is to offer creative alternatives

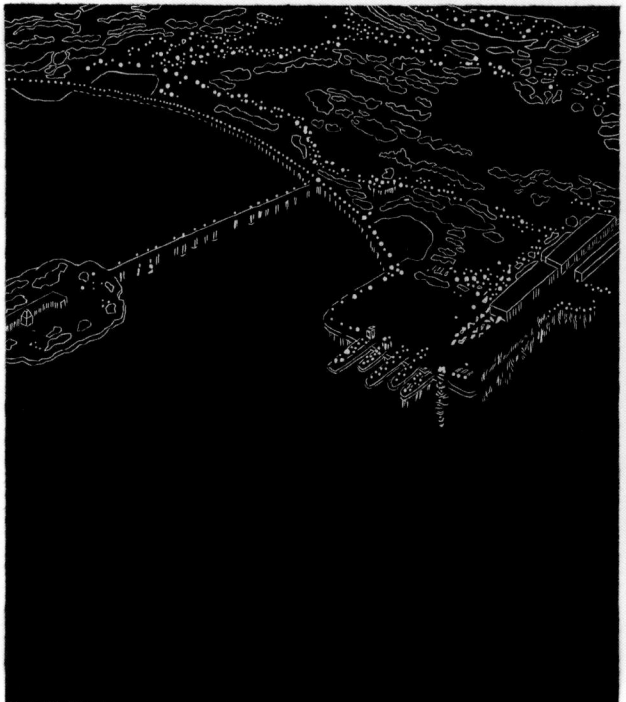
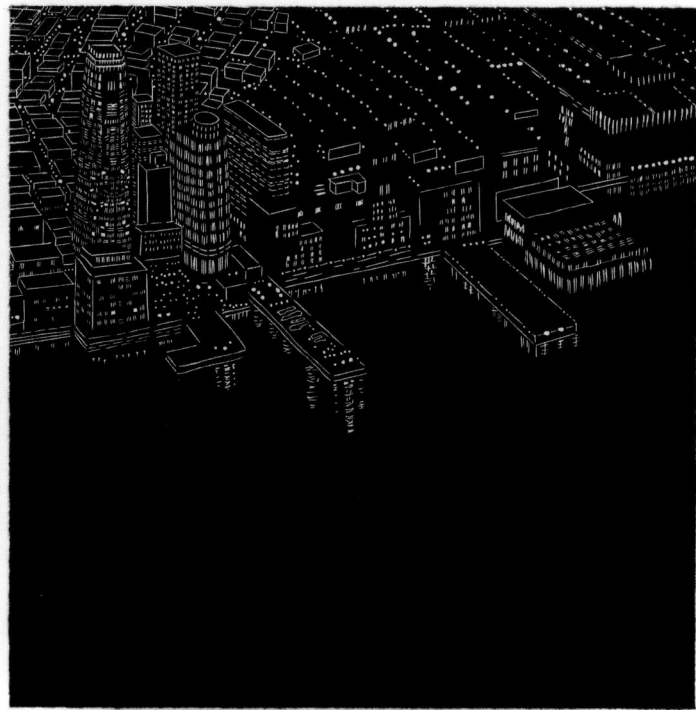

and suggest ways to remake and repair the world."³³ *Shifting Shorelines* commingles riverine art and environmental thought to ask how artists can engage in the ecological repair of habitat. Two artists who pay homage to the Hudson's rewilding futures are Yvonne Jacquette and Marie Lorenz. In Yvonne Jacquette's detailed woodcut *Hudson River Diptych* (fig. 52), based on a view from the 107th floor of the World Trade Center, light and depth unfold across the coastline from the Battery to the Bronx. Jacquette unravels familiar geographies while discovering neglected waterfronts. Lorenz's immersive *Tide and Current Taxi* photographs taken from the mouth of the Hudson River looking up the New York harbor (fig. 64) celebrate the same scene in Frederycks's engraving four hundred years later. Now the city of dreams rising out of Manhattan schist, New York is resilient, vulnerable, sinking. Lorenz's defamiliarizing viewpoint floating upriver from New York Harbor celebrates a renewed sense of the restoration of nature in the Hudson River. Cormorants, Canada geese, and black ducks once again share the Hudson-Raritan estuary habitat with shad, seabass, and oysters.³⁴

Fig. 52. Yvonne Jacquette (1934–2023). *Hudson River Diptych*, 2007. Woodcut, 21 × 40 in. (53.3 × 101.6 cm). Collection of the artist, courtesy of Mary Ryan Gallery, New York

Notes
1. Joel A. Tarr, "Land Use and Environmental Change in the Hudson-Raritan Estuary Region, 1700–1980, with an Addendum to 2018," in Carl A. Zimring and Steven H. Corey, eds., *Coastal Metropolis: Environmental, Histories of Modern New York City* (Pittsburgh: University of Pittsburgh Press, 2021), 10.

2. Robert Grumet, *First Manhattans: A History of the Indians of Greater New York* (Norman: University of Oklahoma Press, 2011), 3.
3. Eric Sanderson, *Manahatta: A Natural History of New York City* (New York: Harry N. Abrams, 2013), 161.
4. Grumet, *First Manhattans*, 3–22.
5. Tom Lewis, *The Hudson: A History* (Harrisonburg: R. R. Donnelly), 37.
6. Jean M. O'Brian, *Firsting and Lasting: Writing Indians Out of Existence in New England* (Minneapolis: University of Minnesota Press, 2010), xxiii–53.
7. Max Liboiron, *Pollution Is Colonialism* (Durham and London: Duke University Press, 2021), 42.
8. Anne-Marie Cantwell and Diana diZerega Wall, *Unearthing Gotham: The Archaeology of New York City* (New Haven: Yale University Press, 2001), 119–48.
9. Jaap Jacobs, *The Colony of New Netherland: A Dutch Settlement in Seventeenth-Century America* (Ithaca and London: Cornell University Press, 2009), 88.
10. Susan L. Glen, *Images of America: Governors Island* (Chicago: Arcadia Publishing, 2006), 9.
11. Gajus Scheltema and Heleen Westerhuijs, eds., *Exploring Historic Dutch New York: New York City, Hudson Valley, New Jersey, and Delaware* (New York: Museum of the City of New York, 2011), 38.
12. Charles Hemstreet, *The Story of Manhattan* (New York: Charles Scribner's Sons, 1901), ch. 3.
13. Michel Trouillot, *Silencing the Past: Power and the Production of History* (London: Beacon Press, 2015), 29–31.
14. Russell Shorto, *The Island at the Center of the World: The Epic Story of Dutch Manhattan and the Forgotten Colony That Shaped America* (New York: Vintage Books, 2005), 49.
15. May Joseph, *Fluid New York: Cosmopolitan Urbanism and the Green Imagination* (Durham: Duke University Press, 2013), 47.
16. NYC Municipal Archives, *Guide to the Records of New Amsterdam, 1647–1862*, https://www.nyc.gov/assets/records/pdf/Dutch-NewAmsterdam_MSS0040_MASTER.pdf.
17. Sarah Zorn, "How Oysters Are Cleaning Up New York's Harbor," *Vox Creative* (September 10, 2018), https://www.eater.com/ad/17841372/oysters-nyc-harbor-restoration.
18. Adrienne Day and John Steele, "The Return of New York Harbor's Oysters," *Nautilus* (April 20, 2022), https://nautil.us/the-return-of-new-york-harbors-oysters-238466/.
19. Robert Tochia, "John Marin," *National Gallery of Art Online Editions* (September 29, 2016), https://purl.org/nga/collection/constituent/2643.
20. Tarr, "Land Use and Environmental Change," 10.
21. "Palmer C. Hayden Papers, 1920–1979," Archives of American Art, Smithsonian Institution, Washington, D.C.
22. "Palmer Hayden, Harlem Renaissance Artist and Beyond," *African American Registry: A Non-Profit Organization*, https://aaregistry.org/story/palmer-hayden-harlem-renaissance-artist-and-beyond/.
23. Steven Lerner, *Sacrifice Zones: The Front Lines of Toxic Chemical Exposure in the United States* (Cambridge: MIT Press, 2012), 18.
24. Carl A. Zimring and Steven H. Corey, *Coastal Metropolis: Environmental Histories of Modern New York City* (Pittsburgh: University of Pittsburgh Press, 2021).
25. Eric W. Sanderson et al., eds., *Prospects for Resilience: Insights from New York City's Jamaica Bay* (Washington, D.C.: Island Press, 2016), 22, 37.
26. United States Environmental Protection Agency, "To Conform with Recent Supreme Court Decision, EPA and Army Amend 'Waters of the United States' Rule," EPA Press Office, August 29, 2023, https://www.epa.gov/newsreleases/conform-recent-supreme

-court-decision-epa-and-army-amend-waters-united-states-rule#:~:text= WASHINGTON%20–%20Today%2C%20the%20U.S.%20Environmental,EPA.

27. May Joseph, *Fluid New York: Cosmopolitan Urbanism and the Green Imagination* (Durham: Duke University Press, 2013), 29.
28. Gerard T. Koeppel, *Water for Gotham: A History* (Princeton: Princeton University Press, 2000), 10.
29. Sanderson, *Manahatta*, 222.
30. John Cronin and Robert F. Kennedy Jr., *The Riverkeepers: Two Activists Fight to Reclaim Our Environment as a Basic Human Right* (New York: Simon & Schuster, 1999), 34.
31. Timothy Morton, *The Ecological Thought* (Cambridge: Harvard University Press, 2010), 38–39.
32. Robin Wall Kimmerer, *Braiding Sweetgrass: Indigenous Wisdom, Scientific Knowledge and the Teaching of Plants* (Minneapolis: Milkweed Editions, 2013), 48.
33. Lisa E. Bloom, *Climate Change and the New Polar Aesthetics: Artists Reimagine the Arctic and Antarctic* (Durham: Duke University Press, 2022), 198.
34. Leslie Day, *Field Guide to the Natural World of New York City* (Baltimore: Johns Hopkins University Press, 2007), 109–10.

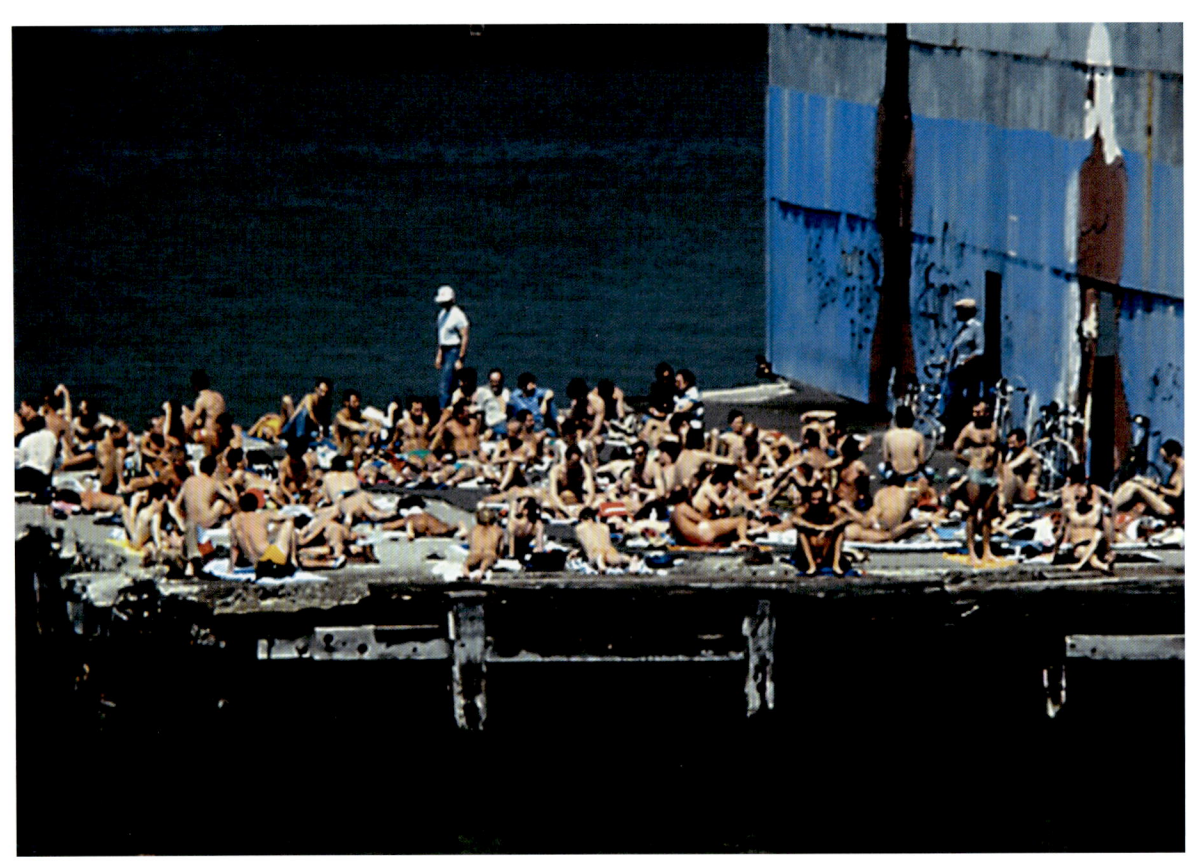

On the Waterfront Again

Jonathan Weinburg

In summer 1978, when I was nineteen and coming out as a gay man, I discovered an extraordinary world along the Manhattan waterfront. I made it the subject of a series of paintings and then, forty years later, the book *Pier Groups: Art and Sex Along the New York Waterfront*. In the 1930s and 1940s a central part of the trade and transportation system of the largest port in the world, the North River piers and the warehouses and terminals that stood on them, from the Battery to Fourteenth Street, had become a series of ruins. The enormous Beaux-Arts buildings on the piers contained halls the size of football fields that had once been used for passenger waiting areas and cargo bays but now were supposedly "abandoned." Manhattan's tightly packed finger piers had become obsolete with the advent of container shipping and the enormous infrastructure the accompanying air and trucking system required.[1]

In a city where space is at a premium, these largely empty structures seemed like an open invitation to anyone who was daring enough, or desperate enough, to make use of them. I was never courageous enough to venture into these buildings at night, but I did like to explore the dilapidated terminal on Pier 46 during the day or sunbathe on the Morton Street open pier. Only a few blocks from some of the most expensive real estate in the West Village, men could be seen lying outside completely naked as Alvin Baltrop, Frank Hallam, and Shelley Seccombe documented in their photographs (figs. 53–55). The burnt-out buildings on the adjacent piers became what the painter Delmas Howe called an "arena for sexual theater,"[2] depicted not only in his own pictures but also in the sexually explicit tableaux created by photographers like Arthur Tress and Leonard Fink.

Lying in the sun or sketching the collapsing structures, I had no idea that only three years before, Gordon Matta-Clark had created his most ambitious work of art in the enormous shed on Pier 52, a few blocks north of where I was sunbathing. Padlocking out the gay men cruising for sex, he and his cohorts cut the great sun-filled arc of *Day's End* into the wall of the cathedral-like hangar. In retrospective, Matta-Clark's act of willfully "enhancing" city-owned property seems like the height of rebellion, but in those years the more shocking work of art was the giant men with engorged penises that

Fig. 53. Frank Hallam (d. 2014). *En Masse Sunners seen from Pier 45*, 1982

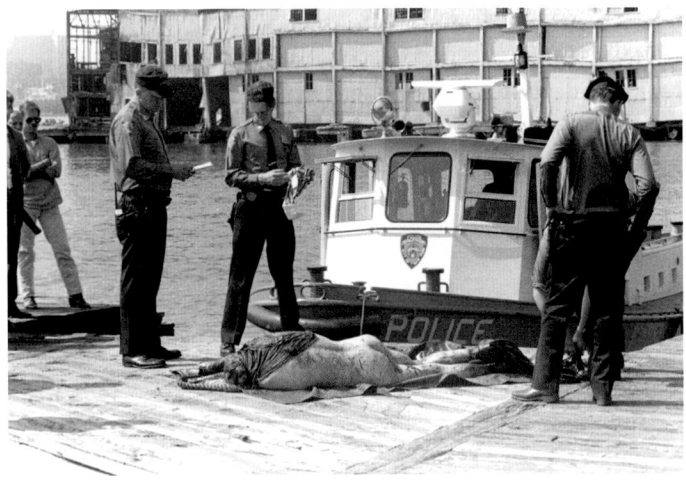 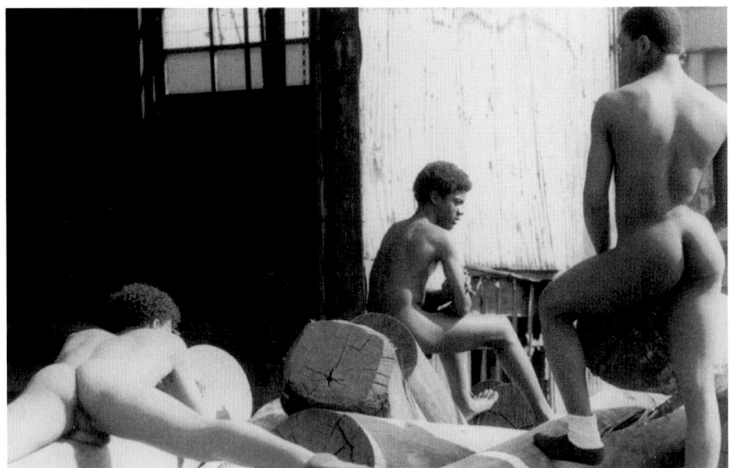

Fig. 54. Alvin Baltrop (1948–2004), *The Piers (police and corpse)*, 1975–86. Gelatin silver print.

Fig. 55. Alvin Baltrop (1948–2004). *The Piers (three men on dock)*, 1975–86. Gelatin silver print.

Tava (born Gustav von Will) painted on the west façade of Pier 46, in full view of the tourist boats that circled Manhattan.

Matta-Clark's aesthetic vandalism was not his first foray into making art on the piers. In 1971 he was one of twenty-seven artists (unfortunately all male), including Vito Acconci, Dan Graham, and William Wegman, who were asked to participate Willoughby Sharp's Pier 18 Project on a part of the waterfront that was eventually filled in to make Battery Park City. Documented by the photographic team of Harry Shunk and János (Jean) Kender, some of the conceptualist works were conceived "in just a few minutes."[3]

Matta-Clark's Pier 18 piece was one of the few works of art created on the piers that directly addressed nature. He transplanted an evergreen tree onto a pile of debris. He then suspended himself upside down in a pose borrowed from the hanged man of the Tarot cards, a figure that symbolizes rebellion and transition.[4]

In contrast, Acconci used the Pier 18 project as an opportunity "to affect, improve, an everyday relationship."[5] In *Security Zone* he was blindfolded. He then asked Lee Jaffe to make sure that he did not fall off the pier as he paced ever closer to the edge.[6] This aspect of *Security Zone*, as well as Acconci's later piece for Pier 17, where he asked people to meet him in the dead of night in order to confess a blackmailable secret, relate to the port's history of crime and corruption depicted in the film *On the Waterfront*.

Like Acconci, the artist David Wojnarowicz risked wandering the interiors of waterfront warehouses at night, sometimes to make art, like the photographs from his *Rimbaud in New York Series*, but also to have sex with men. "Inside, for as far as the eye could see, there was darkness and waving walls of iron, rusting sounds painful and rampant, crashing sounds of glass from remaining windows.... Huge panoramas of factories and water tanks were silhouetted by green roof lights and cars moving down the highway seen only by the red wink of their taillights."[7]

The more I learned about the piers, the more I realized that the idea they were "abandoned" was a misnomer. Reclaimed, they were a hub of creative energies, even as they served as a sanctuary for the homeless and the marginalized until gentrification in the 1990s effectively drove people like the trans-activist Sylvia Rivera, and the LGBTQ youth of color she so ably represented, out of the area. Certainly, radical artists were attracted to the piers because they appeared to be outside or beyond social control. But we should be wary of over-romanticizing this strange moment in the river's history. Baltrop's erotic photograph of three Black youths sunbathing on the waterfront in the nude loses some of its luster when it is put alongside his image of policemen standing over a mangled dead body dredged up from the Hudson—both pictures reflect aspects of the shifting shoreline (figs. 54, 55). Sadly, hard as I try, it is impossible for me to disentangle the story of sexual liberation and experimentation that the piers represent from the arrival of HIV/AIDS, even though the epidemic actually began after most of the pier buildings were demolished. I will always remember how a retired longshoreman scolded me that the ruins I saw as havens of queer liberation and artistic freedom represented to him so many lost jobs and wrecked lives. Indeed, it was Wojnarowicz, whose father was a merchant marine, who suggested how the waterfront in its ruined state was an apt symbol for the collapse of the American Dream: "Soon all this will be picturesque ruins."[8]

Notes
1. For a history of the North River waterfront and its buildings, see Kevin Bone, ed., *The New York Waterfront: Evolution and Building Culture* (New York: Monacelli Press, 1997).
2. Delmas Howe, telephone interview with the author, December 8, 2010.
3. Willoughby Sharp discusses the project in Luca Lo Pinto, "A Portrait of a Transcontinental Cultural Catalyst: A Dialogue with Willoughby Sharp," *Nero* 14 (June–July 2007): 53–55.
4. On Matta-Clark's evocation of the hanged man, see Tina Kukielski, "In the Spirit of the Vegetable: The Early Works of Gordon Matta-Clark (1969–71)," in *Gordon Matta-Clark: You Are the Measure*, ed. Elisabeth Sussman (New York: Whitney Museum of American Art, 2007), 42.
5. Vito Acconci, *Vito Acconci: Diary of a Body, 1969–1973* (Milan and New York: Charta, 2006), 248.
6. Ibid.
7. David Wojnarowicz, *Close to the Knives: A Memoir of Disintegration* (New York: Vintage Books, 1991), 19.
8. Ibid., 24.

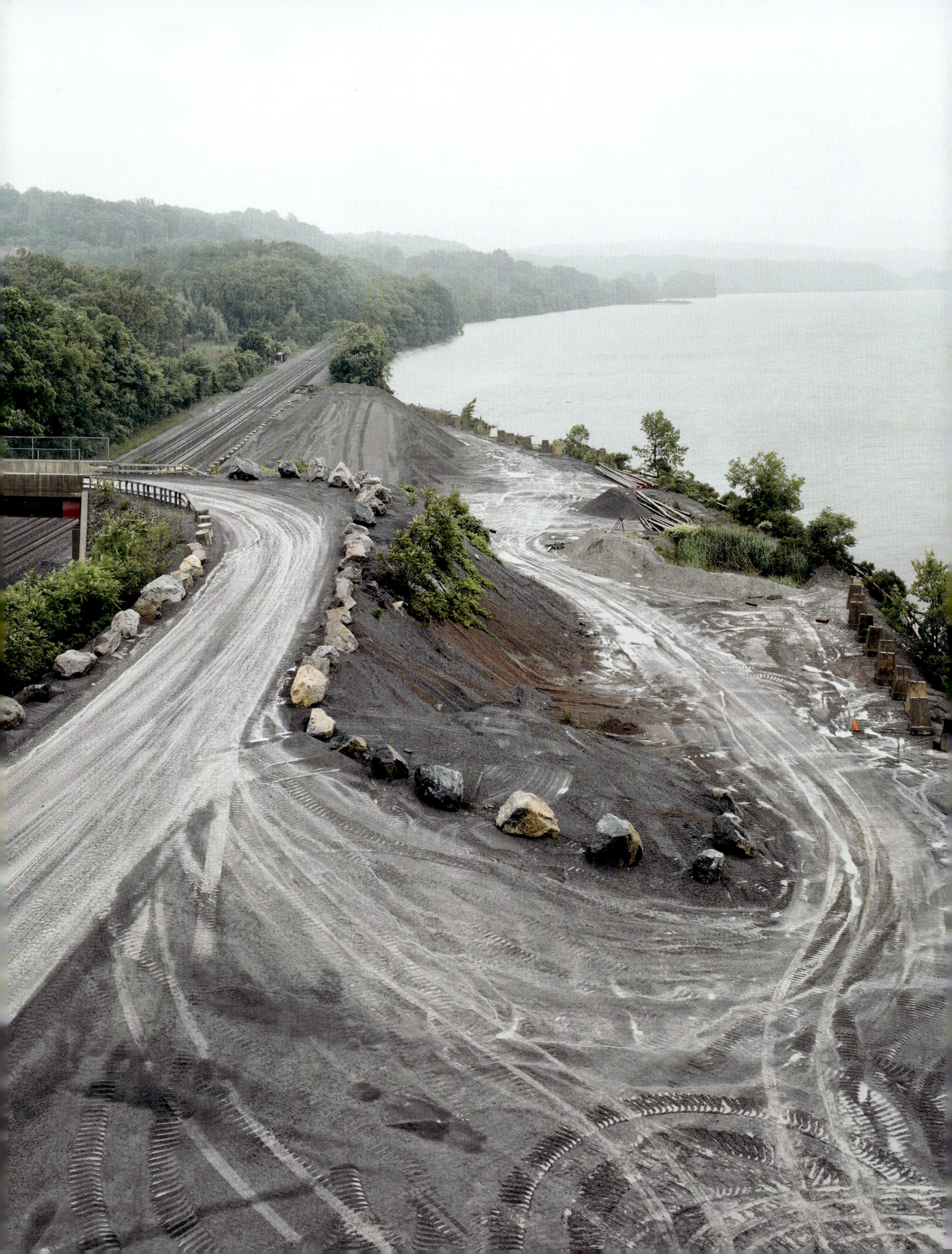

Extraction, Products, and the Environment

Materiality in Contemporary Art of the Hudson River

Betti-Sue Hertz

The mile-high glaciers that covered much of the Hudson Valley twenty thousand years ago contributed to the shape and characteristics of the Hudson River and its environs. The rich deposits of clay left behind along the riverbanks formed the material for ceramics made by the Lenape and the Mohican, who settled in the area during the late Archaic and Woodland periods, six thousand years ago. Archaeological sites have yielded fragments of ceramics in the Hudson Valley dating as early as 1000 BCE crafted by hand, shaped and cut into a variety of styles.[1] From the eighteenth century until the 1940s clay became a valuable resource for the brickmaking industry. By the nineteenth century, Haverstraw and other communities along the river had become major suppliers of bricks for residential and commercial buildings in New York and other burgeoning cities.

For Athena LaTocha, Jean-Marc Superville Sovak, Courtney M. Leonard, An-My Lê, and Marie Lorenz, the qualities of rock, dirt, clay, and minerals are central to their engagement with the Hudson River.[2] While not specifically represented as a body of water, the river is a necessary agent in the geographies and cultural objects that are central to their work. These artists have connected with the riverbed, clay, soil, minerals, pebbles, and rock outcroppings for different reasons, but all have identified an artistic pathway centering on earth and water. Informed by geology (LaTocha), creating a dialogue with brickmaking (Sovak), ceramic oyster jars (Leonard), or a basalt quarry (Lê), or recovering debris from polluted urban rivers (Lorenz), each of these artists focuses on the materiality and the realities of the river and its environs. Aware of the history of nature, the limits of nonrenewable resource extraction, and the production of culturally usable products, they make things that reflect and dialogue with the histories they encounter.

The river is alive. It is a witness to the layers of human and ecological modifications and erasures through neglect or entropy. Undeniably the ecological health of the river and its shores was sacrificed to nineteenth- and twentieth-century notions of progress. When contemporary cultural producers engage with the river, the historical layers of landscape cannot be ignored. Environmental humanities scholar Rob Nixon succinctly states, "Change is a cultural

An-My Lê. *Hudson River* from *Trap Rock* (detail), p. 114

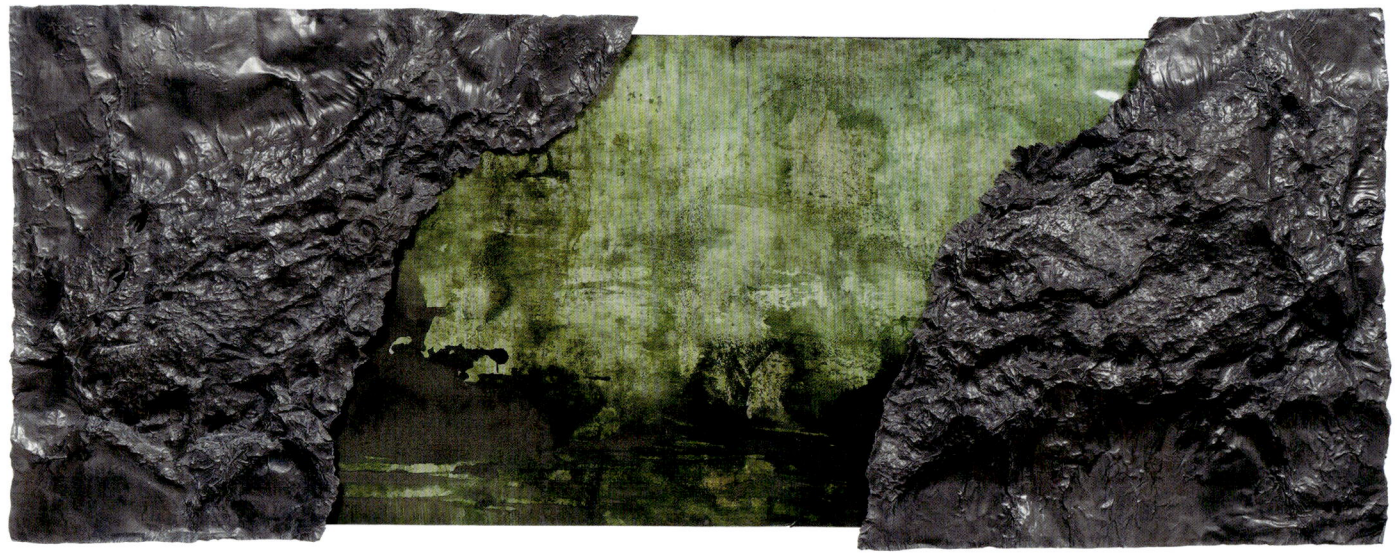

Fig. 56. Athena LaTocha (b. 1969). *The Discovery of Slowness*, 2022. Shellac ink, silt from a Garrison stream, Hudson Highlands mica on paper, lead, steel, 46 × 122 in. (116.8 × 309.8 cm). Antonio Murzi and Diana Morgan Collection

constant but the pace of change is not. Hence the temporal contests over how to sustain, regenerate, exhaust, or obliterate the landscape as a resource become critical."[3] Human time and geological time have a role to play in the construction and deconstruction of geographies. The artists speak to these tensions by uncovering and reimagining the past that is embedded in the present. They reference the material that the river and its environs yields as well as the products that have historically come to define the region, speaking not only to the legacy of earlier inhabitants and later entrepreneurs but also to the origins of the river itself as a byproduct of the last Ice Age. Their projects, and the making of art, cut into Anthropocene narratives and acknowledge the process of human intervention and dependence on nature. The usability of its natural resources, their exploitation, and, ultimately, their depletion follow the rise and fall of the industries that become a source for reuse and commentary by contemporary artists.

The underpinnings of Athena LaTocha's project are rooted in the history of the river, especially the bedrock and rock outcroppings formed in three mountain-building episodes one billion years ago and later relic beds of glacial lakes and deposits of glacial sediment beginning twelve thousand years ago. LaTocha, from Alaska and a member of the Standing Rock Sioux Tribe, engages with the slowness of geological time and space. Eschewing a naive approach to the history of place, she records her findings by incorporating the dirt and imprints of rocks. For mixed-media paintings such as *The Discovery of Slowness* (fig. 56), she pounded lead sheets to make impressions of the surface of rock outcroppings of the Hudson Highlands. The impressed lead sheets flank a painting created through the informal pouring and layering of a

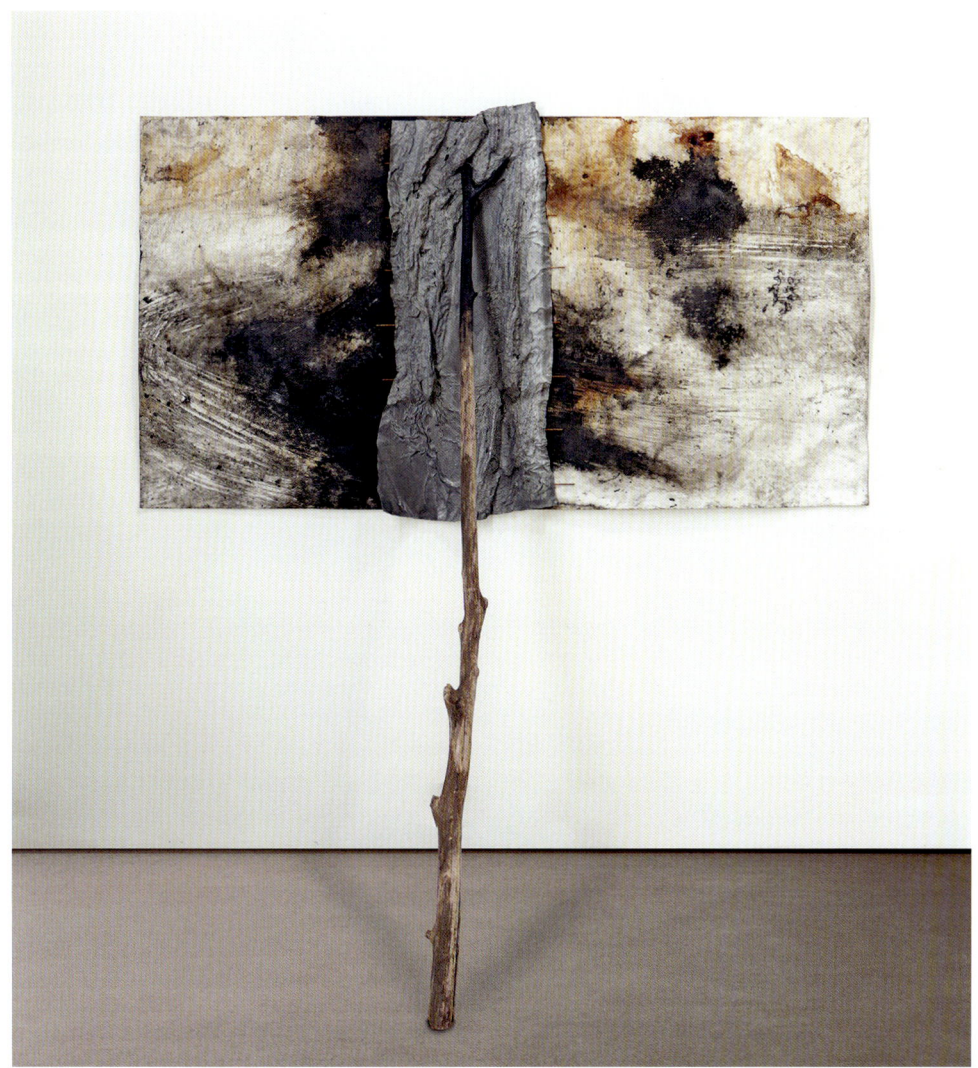

Fig. 57. Athena LaTocha (b. 1969). *Murderer's Creek*, 2018–19. Ink and earth on paper, burned wood, lead, steel, 84 × 84 × 39 in. (213.4 × 213.4 × 99.1 cm). Private collection

mixture of shellac ink, silt from a Garrison stream, and locally sourced mica. With these procedures, LaTocha achieves the characteristics of a landscape painting. The traditional composition of water, land, foliage, and sky collapses into scumbly fields of color heavily tinted green. Her landscapes edge toward the visual qualities of chemicals, toxicity, and ruin while reaching for sublime beauty. She has also created works with materials retrieved closer to the shoreline such as *Murderer's Creek* (fig. 57) and *Winterkill* (2018–19), which include worn branches she pulled out of the Hudson River adjacent to Wave Hill, a cultural center and park in the Riverdale neighborhood of the Bronx.

In other LaTocha paintings the horizontal or ascending/descending planes of the ground become vertical. This action is at a remove from the conventions of Earthworks artists of the 1970s such as Robert Smithson and Nancy Holt. Yet, conceptually LaTocha's work aligns with that of Smithson,

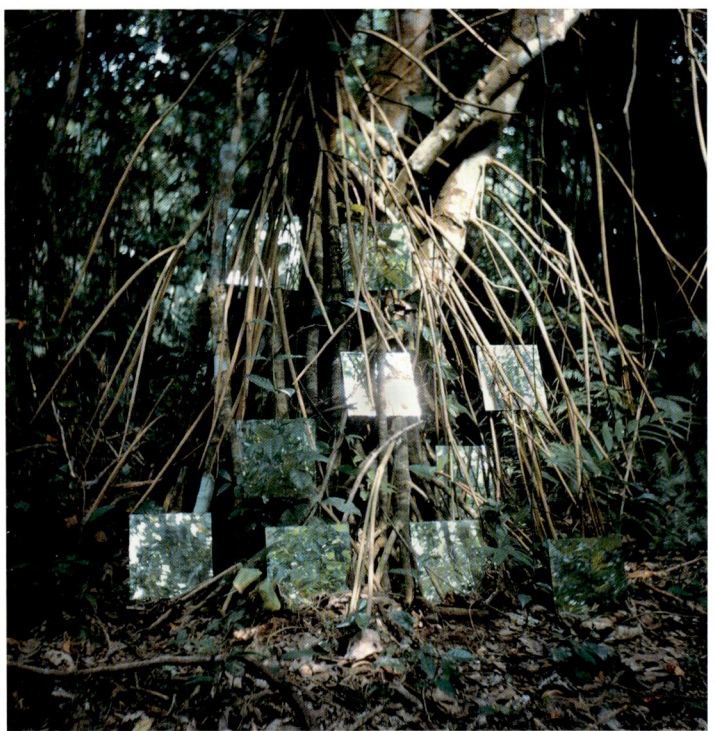

Fig. 58. Robert Smithson (1938–1973). *Yucatan Mirror Displacements 1–9*, 1969. One of nine chromogenic prints from chromogenic slides (126 format), 24 × 24 in. (61 × 61 cm). Solomon R. Guggenheim Museum, New York, Purchased with funds contributed by the Photography Committee, and with funds contributed by the International Director's Council and Executive Committee Members: Edythe Broad, Henry Buhl, Elaine Terner Cooper, Linda Fischbach, Ronnie Heyman, Dakis Joannou, Cindy Johnson, Barbara Lane, Linda Macklowe, Brian McIver, Peter Norton Foundation, Willem Peppler, Denise Rich, Rachel Rudin, David Teiger, Ginny Williams, and Elliot K. Wolk, 1999

who positioned the viewer to engage with the earth in installations of dislocation such as *Yucatan Mirror Displacements, 1–9* (fig. 58), exposing how landscape is constructed, inclusive of extraction and the products it yields. Smithson's anti-monuments might lead us to LaTocha's paintings. She makes spatial sense of the real, if somewhat already disturbed, natural material while retaining representational, or at least illusory, associations with landscape. She subverts and reinvents the sublime aspects of the Hudson River School painters who created works that reached the pinnacle of artistic achievement while avoiding denial of the environmentally destructive outcomes of the region's lucrative industries.

Artists who do something with Hudson River phenomena and histories are a testament to the natural world as a giving space for artistic invention. When natural resources are turned into products one step removed from their material origins, it opens up culturally specific approaches to this materiality. The structure of capital will mean that ultimately, if manufacturing goes unchecked, resources will be depleted and the dependent sites abandoned. Such was the case with the brickmaking factories, the quarries, oyster farming, and many other industries that flourished along the Hudson River in the eighteenth and nineteenth centuries. The artists are picking up the pieces and trying to make sense of them.

Jean-Marc Superville Sovak's research-based work focuses on the lives of Blacks in the Hudson Valley. Connecting the clay deposits that supported brickmaking, Sovak highlights the Underground Railroad (1810–61) and the Reconstruction Era (1865–77) as critical for formerly enslaved and free Blacks. For Sovak, a Canadian artist of Trinidadian and Czech descent living in Wallkill, New York, retelling local histories creates a legacy narrative for the multigenerational residents of the Catskills and the Hudson Valley. Harvesting along the shores of the river, he salvages rejected bricks from the Empire company's brickyard (ca. 1910, closed 1940) that, like the bricks from the Hutton yards (ca. 1870–1980) and others, are still strewn along the riverbanks.[4] Blacks, usually from the American South, and Italian and Irish immigrants were among the workers recruited to labor in the brickyards in the late nineteenth and early twentieth centuries.[5] He intervenes in images and narratives that have represented the region by honoring the lives of

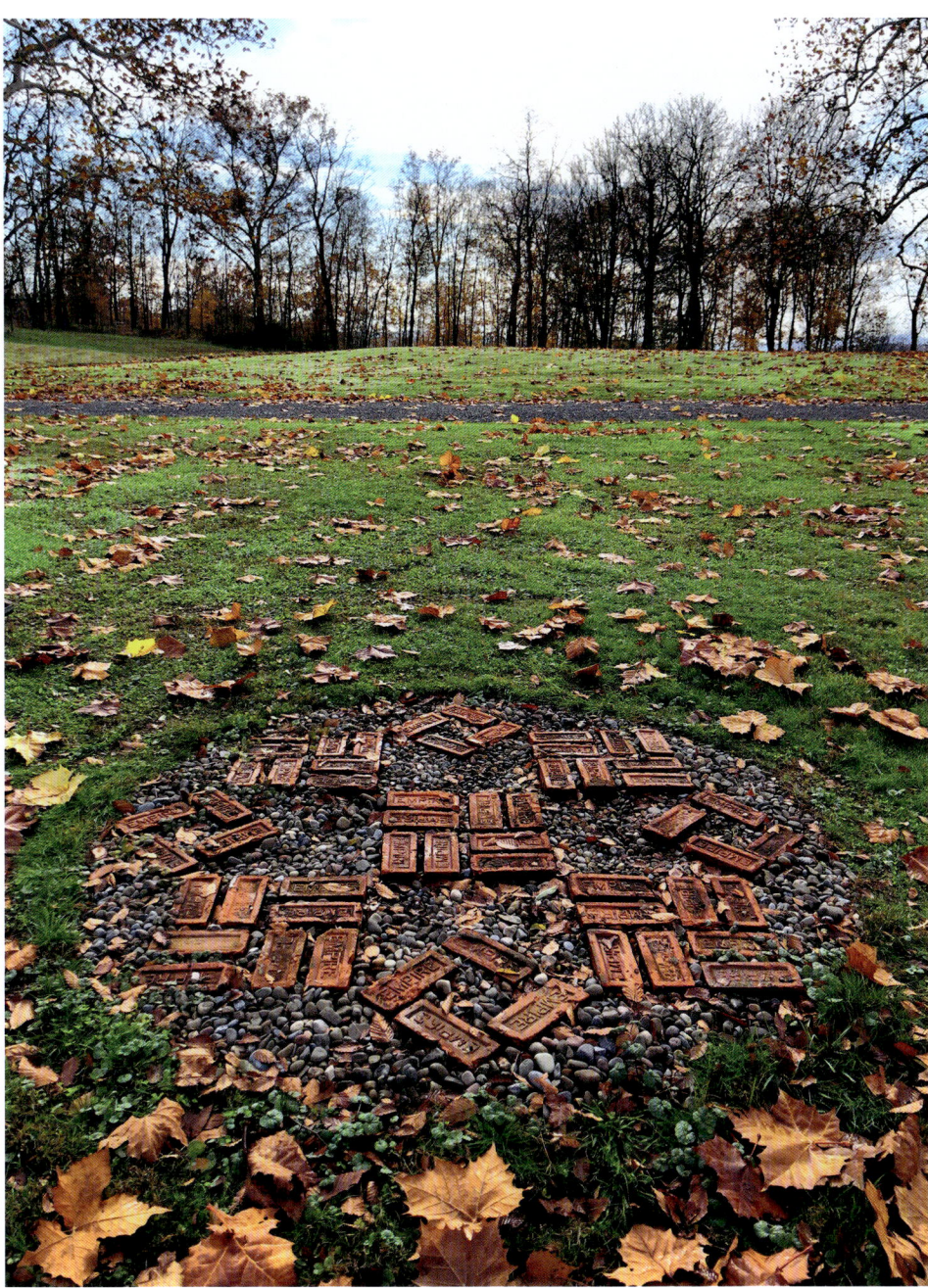

Fig. 59. Jean-Marc Superville Sovak (b. 1976). *Freeborn Trails (North Star)*, 2021. Courtesy of the artist

Black workers whose labor benefited white elites. Sovak reconstructs history near sites of the brick-factory ruins to awaken the temporalities of place and of past lives. Through this effort he resurfaces local histories of labor, the abolitionist movement, and artistic cultures of the nineteenth-century Hudson Valley.

In the *Freeborn Trails* series (fig. 59), Sovak installed reclaimed bricks with their EMPIRE stamp surrounded by river stones into heritage-quilt

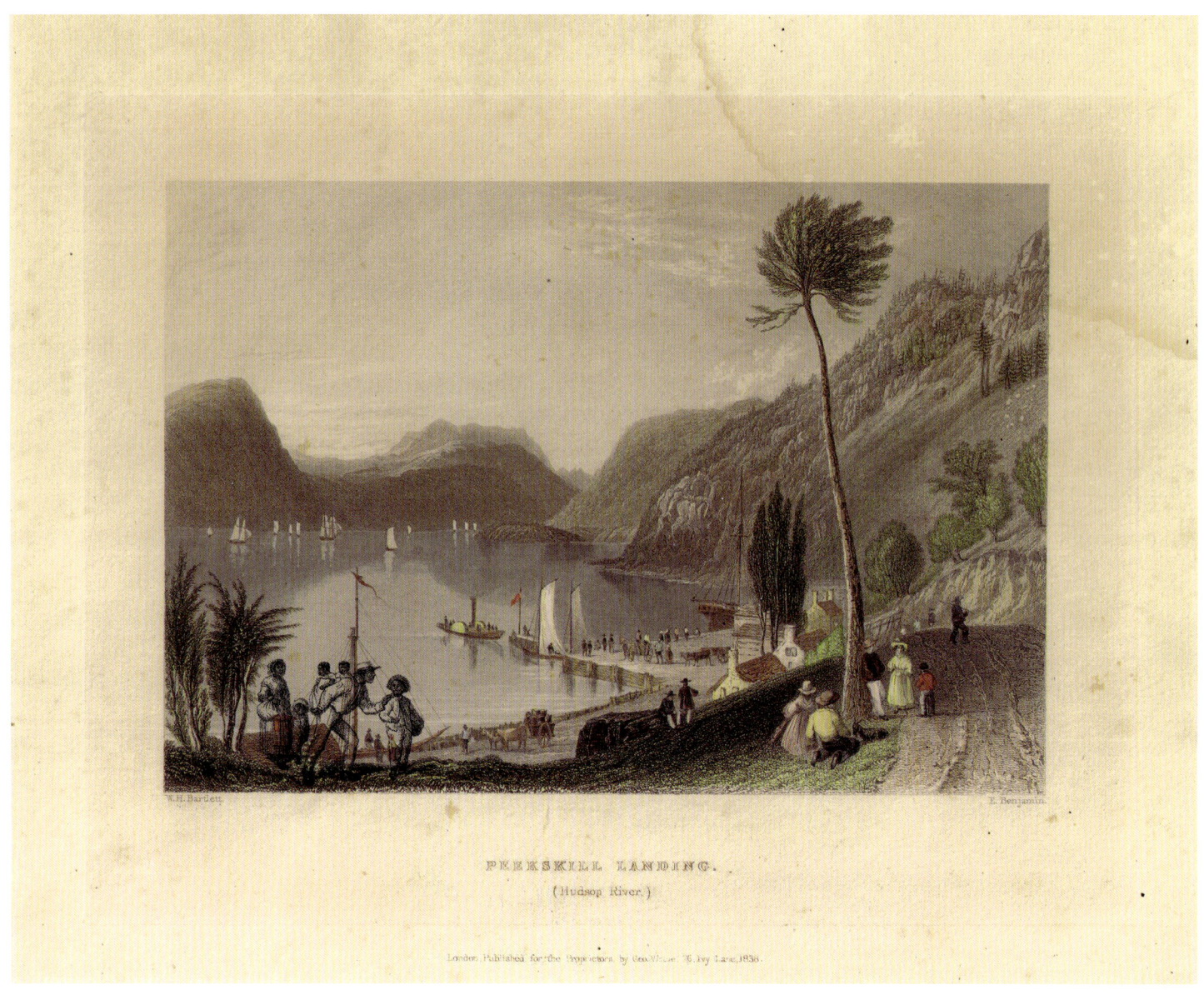

Fig. 60. Jean-Marc Superville Sovak (b. 1976). *a-Historical Landscape: Peekskill/Runaways*, 2019. Monoprint on found engraving, 8 × 10 in. (20.3 × 25.4 cm). Collection of the artist

patterns by placing the baked clay back into the land and embedding the bricks into six discrete sites at the Wilderstein Estate in Rhinebeck. A house museum that includes property originally purchased by the abolitionist Methodist minister Reverend Freeborn Garretson, it was also a stop on the Underground Railroad. The series includes several designs, among them the "North Star" pattern.[6] In another body of work, the *a-Historical Landscape* series (fig. 60), the artist superimposed figures from anti-slavery almanacs and abolitionist tracts into popular engravings of Hudson River scenes by the British illustrator William Henry Bartlett (1809–1854).

Some contemporary artists look to the same resources that had been an economic driver for previous generations. Courtney M. Leonard fashioned a newly commissioned sculpture as a tribute to oyster farming along the Hudson, with a focus on the handmade salt-glaze ceramic jars for storing and preserving pickled or fried Manhattan oysters. Leonard, a member of the Shinnecock Nation based on Eastern Long Island, often makes works related to industries that have supported the livelihoods of Native peoples there. Much of her practice is centered on the whaling trade, colonial interruption of Native practices on Long Island, and Native whaling culture. Committed to eco-activism, she focuses on ocean preservation and the impact of nets, plastics, and pollutants on ocean mammals. Leonard's work is meticulously crafted using traditional Native American materials such as sparkling micaceous clay, often employed for cooking vessels. The ceramics in her *Breach: Logbook* series (fig. 61) reference whale teeth, scrimshaw, and Indigenous

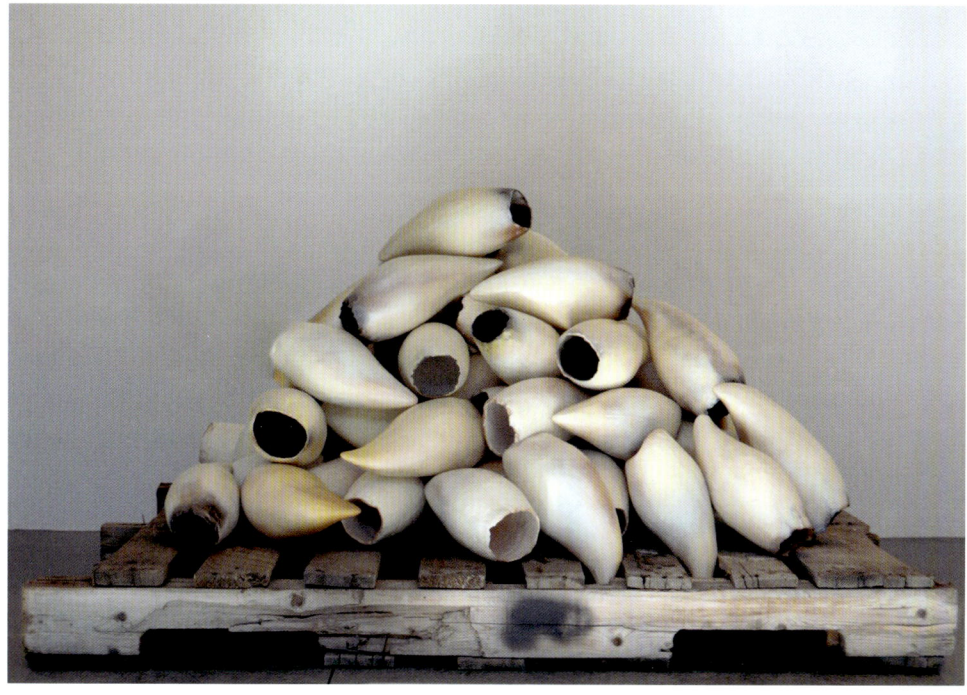

Fig 61. Courtney M. Leonard (b. 1980). *Breach #2*, 2016. Ceramic and wood pallet, approx. 36 × 48 × 36 in. (91.4 × 121.9 × 91.4 cm). Collection of the artist

EXTRACTION, PRODUCTS, AND THE ENVIRONMENT

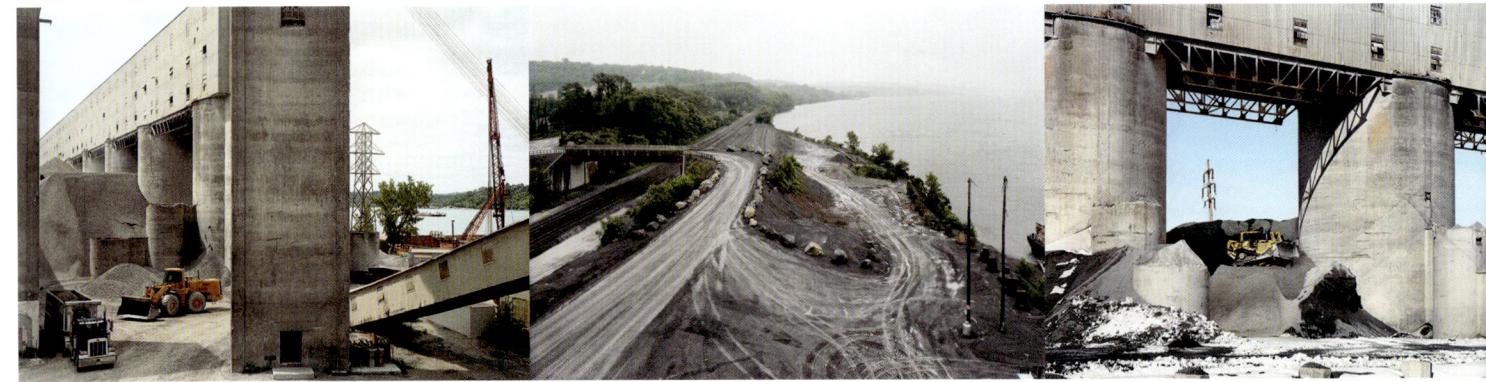

Fig. 62. An-My Lê (b. 1960). From left to right: *Truck Load Out (River)*, *Hudson River, Mill Silos III (Backhoe)*, *Beach Parts*, *Excavator*, and *Fixed Stackers* from *Trap Rock*, 2006–7. Pigment prints, each 30 × 42 in. (76.2 × 106.7 cm). Collection of the artist, courtesy of Marian Goodman Gallery, New York

fishing baskets. They relate to the availability and sustainability of water and to the cultures and traditions that have been subject to colonial violence. With this background she brings expertise in the complexity of marine life as sustenance in the Hudson River.

Long before colonial contact the Lenape were aware of ocean mammals (whales and seals) and fish of all sizes and descriptions.[7] Oyster reefs native to the New York Harbor were an important part of marine and food ecologies. As American Studies scholar Ayasha Guerin notes, "But the oysters of New York harbour would share the same fate of the right whales of New England. New York's shift from a communitarian-based economic order to a market society and money-based economy meant the oyster was commodified and the native beds were quickly exhausted at the turn of the 19th Century."[8]

For her new work, Leonard looks to the river's past, oystering, and the ceramic jars that became central to that industry. Marine culinary cultures provided a livelihood for the Shinnecock and the Lenape, free Blacks, and white oystermen.[9] Of note are the occupations that, at the time, had a very low status and were often owned by free Blacks. The master potter Thomas Commeraw supplied jars (fig. 11) to business owners like Henry Scott; both African American men were leaders among the oystermen of lower Manhattan. The cylindrical jars and their unique histories are a starting point for Leonard's ceramic work in the exhibition.

While the public's focus is on global removal of fossil fuels, mining of precious metals, and climate change, there are other ways extraction strips the earth and its inhabitants of healthy life. Industries that developed and sustained communities along the Hudson River—oyster harvesting and farming, brickmaking, mining, fishing, forging, agriculture, logging—have all been impacted by the overuse and depletion of natural resources. Artists awakening awareness of these pasts and forging environmental statements through their evidentiary strategies reach back to these rich histories. Woven into

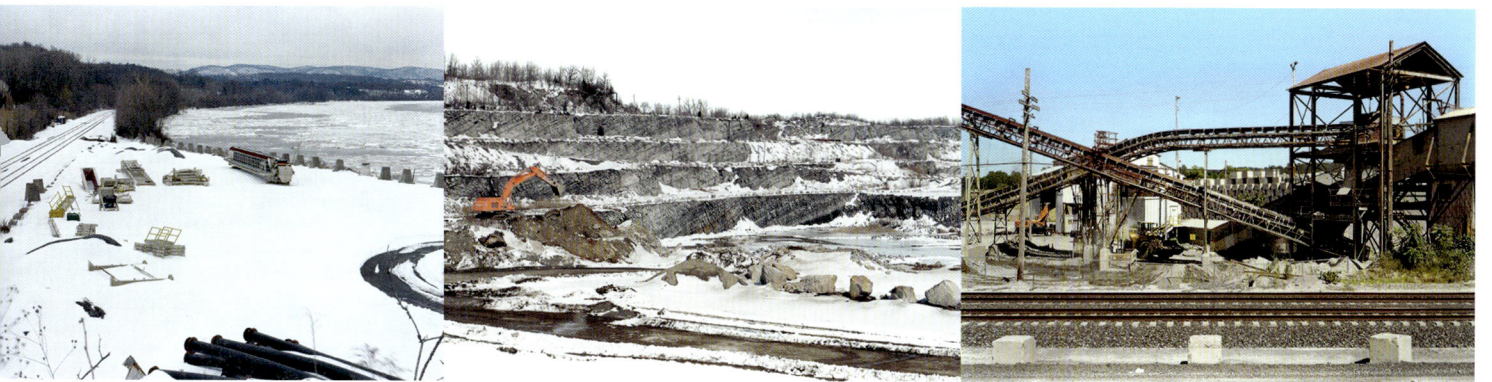

their poetic activism is respect for the makers of commodities like bricks and oyster jars, ironworks, and even parks and roads who are employed to manufacture products out of the natural resources and, as a by-product, reshape or degrade the land and water ecology. They are allies in spirit with environmental groups such as Riverkeeper and Scenic Hudson and local historical societies throughout the region.

The landscape tradition collapses into representations of extraction in An-My Lê's DIA: Beacon commission *Trap Rock* series (fig. 62), twenty-two photographs of a basalt quarry formerly owned by the New York Trap Rock Company. Located outside Poughkeepsie, the quarry opened in 1929. The basalt often reentered the landscape as processed rock used for railroad beds and for embankments to protect coastlines. It thus enabled connectivity between cities and towns, the ability to transport goods, and the evolution of cities.

Originally from Vietnam, Lê brings a sensibility to her work that relies on both observation and the theatrical possibilities of her subject. Her choice to document this quarry using a pictorial strategy that features vastness nods to the Hudson River School painters as well as to Smithson's sensitivity to the ecological problems caused by extraction industries. Lê's *Trap Rock* series is another example of how a contemporary narrative, while acknowledging the past, foregrounds a changed and ecologically damaged present.[10] She focuses on the machinery and on the incision of the earth and the disruption of local ecologies. The industrial scale of the extraction is a dramatic window onto how mining carves terraces into the landscape. Water is a necessary agent for digging and crushing the rock. The presence of workers is primarily made visible through the equipment and the result of their labor. Nature, nevertheless, survived despite the mine.

Another series by Lê that focuses on disturbed landscapes is *29 Palms* (fig. 63). Here, the photographer staged battles at the site in the California

EXTRACTION, PRODUCTS, AND THE ENVIRONMENT

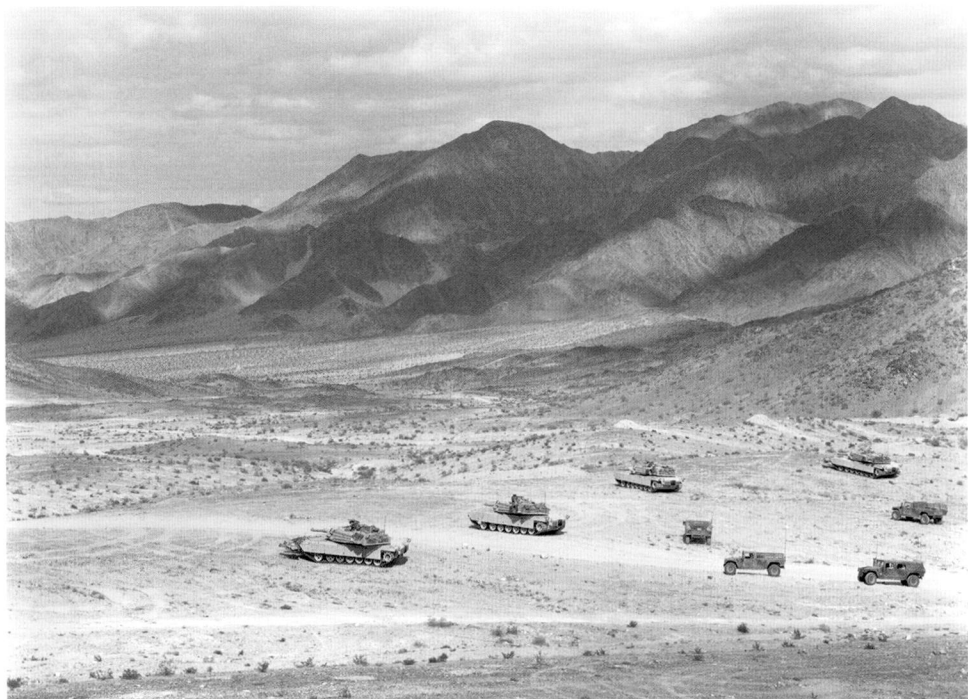

Fig. 63. An-My Lê (b. 1960). *Mechanized Assault,* from *29 Palms,* 2003–4. Gelatin silver print, 37½ × 26 in. (95.2 × 66.0 cm). Museum of Contemporary Photography at Columbia College Chicago, Museum Purchase (2012.51)

desert where U.S. Marines train for battle prior to deployment. These images reference, but diverge from, nineteenth-century landscape photographers of the American West such as William Henry Jackson,[11] Carlton E. Watkins, and Timothy O'Sullivan, whose work is often compared to the majestic scale of Hudson River School painters like Thomas Cole and Frederic Church.

Rock was important for New York City's built environment. The small businesses along the shores of the Hudson River were the suppliers of materials in service to the Industrial Revolution and contributed to this notion of progress by literally accessing the ground that supports the natural and the built environments. One can imagine that while these businesses were bending nature to the production of products, they also romanticized the beauty of the landscape they inhabited.

Industry creates its own waste, both in making the thing it makes and in the post-consumer universe of trash. Marie Lorenz, a white artist who hails from a military family, is taking up another part of this narrative. Her fascination with the physical power of waterways and their vulnerability when subject to human neglect shapes her artistic project. She takes notice and then revives the leftovers of human consumption. Ebb and flow have shaped her practice as well. Lorenz has learned the ways of the New York Bay well enough to have made many successful runs in her *Tide and Current Taxi* project (fig. 64). She takes people across the bay and other locations in a small hand-crafted rowboat, propelled only by oars in a

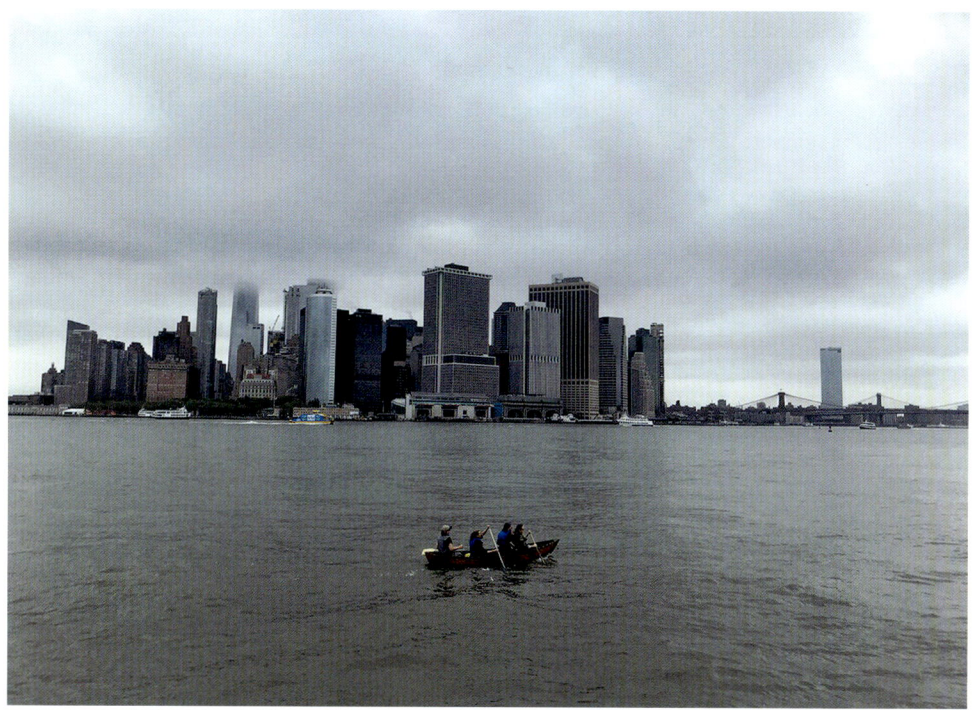

Fig. 64. Marie Lorenz (b. 1973). *Tide and Current Taxi*, 2005–present

collaboration with the tidal rhythms of water. Recent floating sculptures made from natural and consumer materials temporarily bob up and down with the tides. In *Floating Instruments* (2022) she added a sound component to her assemblage, with homemade instruments fashioned to be played exclusively by the wind. Lorenz respectfully exposes the flotsam and jetsam at the intersection of the vagaries of weather, water, and detritus. For *Shifting Shorelines*, Lorenz created a sculptural installation of an imaginary deconstructed, decayed Jet Ski (fig. 65) reminiscent of the bones of large mammals that wash up on the shores of contaminated waterways. The piece is her commentary on the environmental insensitivity of these recreational watercraft even though she admits an enthrallment with the way they zip around he Hudson River.

Industrial production, especially excavation and its by-products as well as the debris that litters the waterways, comprises an ecological language of these artists' creative work. At the intersection of the natural and the cultural spheres, one erodes the other in a complex relationship until one potentially becomes the other. Whether it is the reshaping of the marshes to make way for the nineteenth-century brickmakers or the reordering of the shoreline to make way for parks of landfill, the shoreline is an active space. There are now efforts at reclamation and recovery while retaining a memory of past industry at sites such as the Haverstraw Brick Museum and the Scenic Hudson West Point Foundry Preserve in Cold Spring.

EXTRACTION, PRODUCTS, AND THE ENVIRONMENT

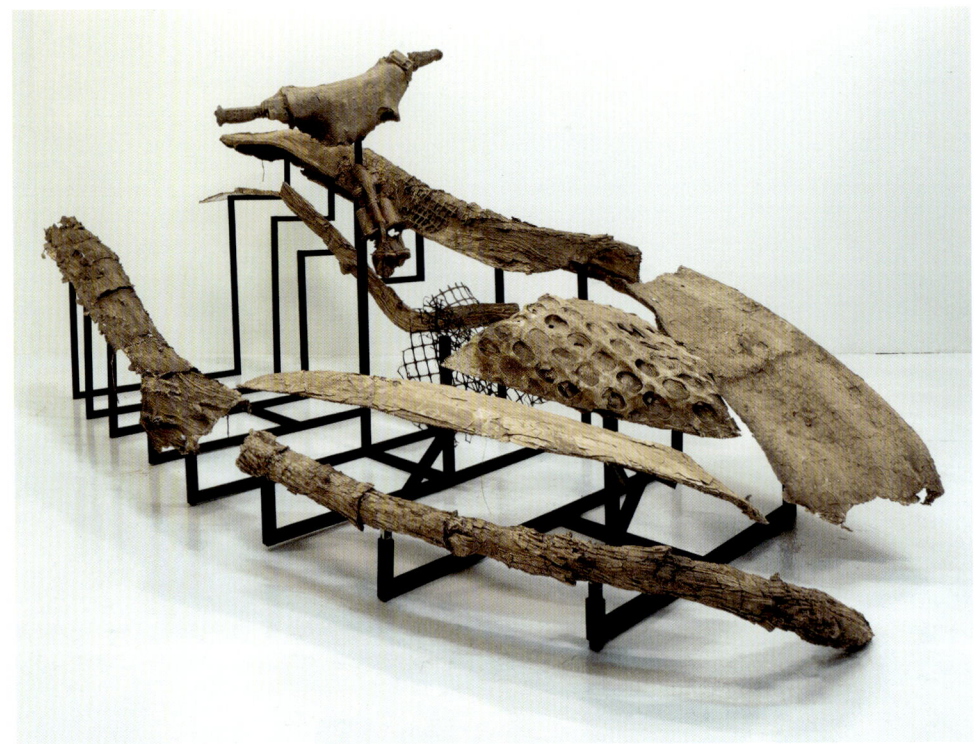

Fig. 65. Marie Lorenz, *Jet Ski*, 2024. Steel, gypsum cement, 60 × 54 × 123 in. (152.4 × 137.2 × 312.4 cm). Collection of the artist

The current generation of artists navigating their relationship with the river has the distinct advantage of grasping the fragility of its health within the larger discourse on climate change. They share a language for understanding how human activity affects the natural resources on which it depends. By manifesting a materials-based perspective of the Hudson River, its cultural and labor history, LaTocha, Leonard, Sovak, Lê, and Lorenz deepen our understanding of the complexities of the earth's offerings and the toll cultural activities have upon them. Their representations of place foreground the importance of demise and critique as well as the beauty of retaining a sensorial interaction with the Hudson River's geological and aquatic histories and its service to the products and commodities yielded from it.

Notes

1. Pottery attributed to the Proto-Munsee bands along the northern Hudson—Aquakanonk, Tappan, and Pompton—generally has elongated bodies with medium to high collars and triangular geometric patterns in complicated interrelationships. Finds attributed to Proto-Unami bands in the south—Sanksikan, Navaskin, and Marites—are usually collarless and shaped like a hornet's nest. Collarless, egg-shaped, or globular vessels with flaring rims and bodies roughened by cord markings and linear designs were common at the Owasco site in upper New York. See Herbert C. Kraft and R. Alan Mounier, "The Late Woodland Period in New Jersey (ca. A.D. 1000–1600),"; https://www.nj.gov/dep/hpo/1identify/pg_139_LateWdlnd PeriodNJKraft_Mounier.pdf.

2. Other contemporary artists in the exhibition engage with the cycles of production and ruin but from perspectives other than geological. For example, Every Ocean Hughes's photographs of weathering pilings in *The Piers Untitled* series (2009–23); Gordon Matta-Clark's photographs, videos, and performances of a decaying industrial pier building in *Day's End* (1975); and David Hammons's *Day's End* public art project (2014–21) on the same site as Matta-Clark's images. These artists treat the piers on Manhattan's West Side as sites for recovering the passage of history, time, and loss.
3. "More than material wealth is here at stake: imposed official landscapes typically discount spiritualized vernacular landscapes, serving webs of accumulated cultural meaning and treating the landscape as if it were uninhabited by the living, the unborn, and the animate deceased." Rob Nixon, *Slow Violence and the Environmentalism of the Poor* (Cambridge: Harvard University Press, 2011), 17.
4. At the beginning of the twentieth century, 130 brick manufacturers employed seven to eight thousand workers on the Hudson River. At its peak it was the largest brick-making region in the world. By 2002 the industry ceased to exist. See George V. Hutton, *The Great Hudson River Brick Industry: Commemorating Three and a Half Centuries of Brickmaking* (New York: Purple Mountain Press, 2003).
5. At the height of Hudson Valley brickmaking, 60 percent of the workforce was Black; see "Bowline Brickyards Ruins," African American Society of Rockland County, https://aahsmuseum.org/dt_portfolios/bowline-brickyard-ruins/. The Haverstraw Brick Museum is a wonderful resource for learning more about this history: https://www.haverstrawbrickmuseum.org/from-thearchives/bricktalk-labor-in-the-brickyards-acknowledging-the-untold-stories.
6. According to Sovak, "The North Star is perhaps the most recognizable of the Freedom quilt patterns because it is both descriptive and prescriptive of the direction those who were self-emancipating from slavery needed to follow. . . . This pattern is a reminder that like the North Star, freedom is always directional and aspirational," http://www.supervillesovak.com/uploads/3/0/8/2/30820225/north_star.m4a.
7. "When a certain small tree blossoms white in the spring, the *schawanammek*—shad—have started to swarm upriver to spawn. Four hundred foot long nets woven out of tree bark fiber get staked in the shallows; wooden markers carved into the faces of *manëtuwàk*—spirits—show where they're set and call the fish in. Then, so many shad and stripers and eel bulge the nets that the wooden faces bob up and dance." Daniel Wolf, "A Glimpse of the Lenape: The Night Before," https://www.riverkeeper.org/hudson-river/hudson-river-journey/the-first-people-of-the-river/a-glimpse-of-the-lenape/page-2/.
8. Ayasha Guerin, "Underground and at Sea: Oysters and Black Marine Entanglements in New York's Zone-A," *Shima* 13, no. 2 (2019), https://shimajournal.org/issues/v13n2/06.-Guerin-Shima-v13n2.pdf. The oysters, once plentiful, were almost completely depleted, and the industry collapsed because of overharvesting and water pollution about 1920.
9. The oyster jars often traveled on trade ships headed to sugar-cane producing countries as gifts to captains and plantation owners. As a postscript, the Billion Oyster Project founded in 2014 is actively and successfully reseeding the oyster beds in New York Harbor. See https://www.billionoysterproject.org/.
10. A reference to Robert Smithson's *Rocks and Mirror Square II* (1971) may be useful here as it involved the removal of basalt from a trap-rock quarry in Paterson, New Jersey, which was installed in a gallery. Lê's photographs offer a view of the quarry that relies on placement in the exhibition space, this time through image rather than thing.
11. Jackson also photographed the Hudson River; see *Spuyten Duyvil, First View of the Hudson* (1890s) and *Palisades of the Hudson* (ca. 1900).

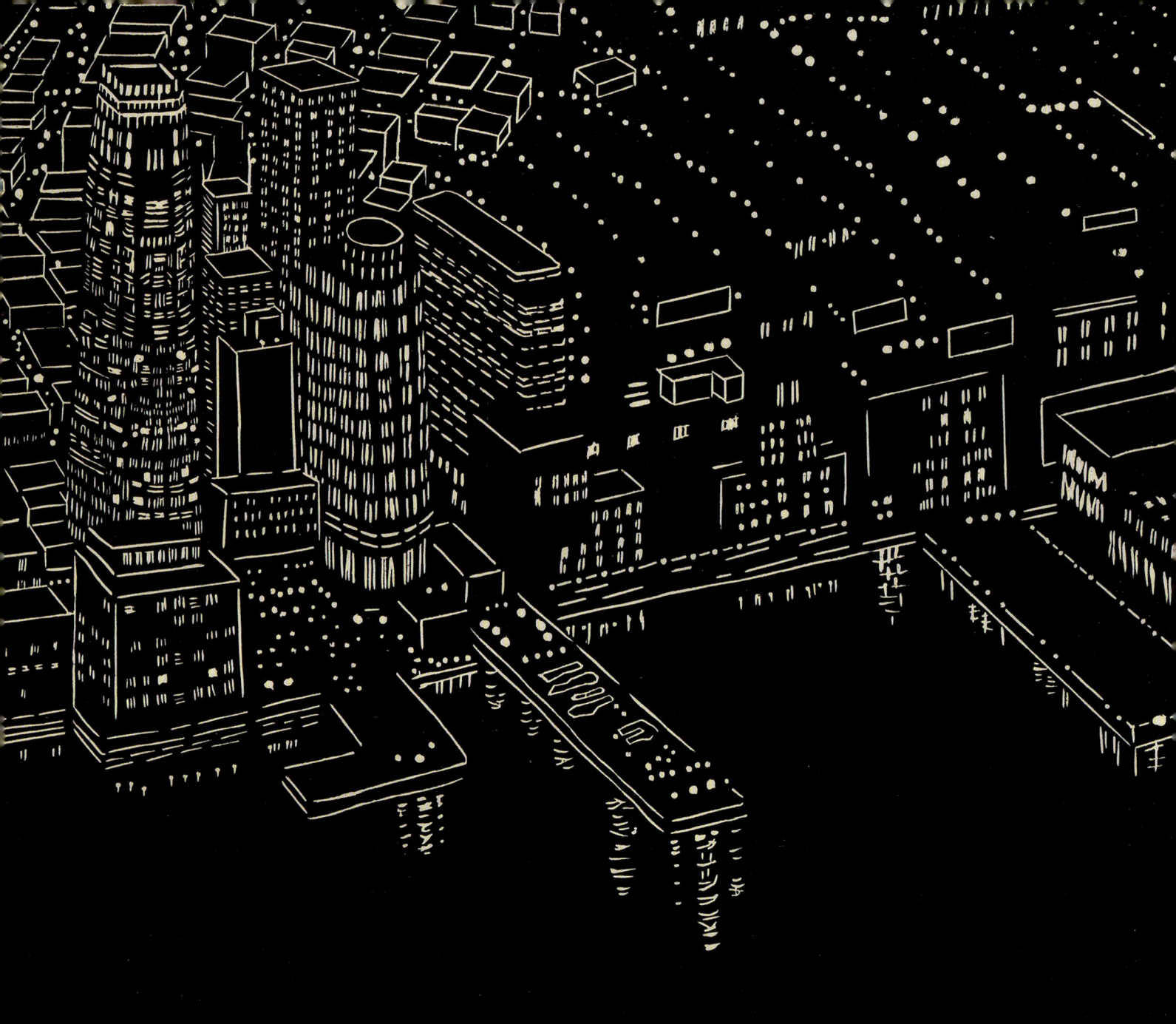

Artists' Roundtable

Moderator
Betti-Sue Hertz, Director and Chief Curator, Wallach Art Gallery

Artist Participants
Athena LaTocha (Hunkpapa Lakota/Ojibwa)
Jean-Marc Superville Sovak
Courtney M. Leonard (Shinnecock)
Marie Lorenz

B.H. You might know each other's work, but maybe you haven't met. Athena, do you want to start us off?

A.L. My name is Athena LaTocha. I've been in New York for about twenty, twenty-five, twenty-seven years. The idea of landscape is very close to me, having been born and raised in Alaska, where I was immersed in the wilderness—hunting, fishing, collecting, berry picking, and other outdoor activities. How do we reconcile human impact on land? How do we work with those ideas moving forward as a culture?

C.L. My name is Courtney Michelle Leonard. I'm from the Shinnecock Nation of Long Island, New York. I'm joining you today from Dakota Territory in Minnesota. My work has a lot to do with being people of the water and people of the land. I look at coastal erosion and high tide and our different knowledges about water based on the actual daily lived experience of our environmental impact. It's also tied historically to my family and our nation's connection to the path of the whale and our relationship to the whales. The whale exists as a metaphor in my series *Breach* in terms of experiences of indigeneity, but not specifically a conglomerate of Indigenous experiences. I'm working on a project right now in relationship to northern right whales, orcas, and the different pods that used to come out our way historically. I also actively connect with people from other water communities to learn about larger issues having to do with whales. A lot of people care about whales, not too many people care about small Indigenous coastal communities. So, it's important to amplify our issues, especially when there are people who tend to homogenize Indigenous communities when it comes to the topic of water and conservation.

Yvonne Jacquette. *Hudson River Diptych* (detail), p. 99

J.S. Hi, I'm Jean-Marc. I'm a recent arrival to the Hudson Valley where I went to college and where my daughter was born, so it's as close to anything I've called home. I focus on the area in the vicinity of the towns of Beacon, Newburgh, New Paltz, and Poughkeepsie. The more I learn, it seems there is history here under every surface. I'm especially interested in learning more about the area's first Africans who were bought—enslaved—until arguably 1827. I live ten minutes from where Sojourner Truth was born, in a town called Swartekill—present-day Rifton. And if you drive there, there's not a sign, not a plaque, nothing about her. I'm dedicated to making that history visible.

M.L. Hi, everyone. I'm Marie Lorenz. I've been in Brooklyn for over twenty years, and my work has always been about water and tide navigation. New York City is mostly water, and I love thinking about that in relation to the city.

B.H. How are the waterways and shorelines reflected in your work? Some of you work directly with water, and some of you are working on the land nearby. All of you have had some connection to the waterways around New York City or farther up the Hudson. Marie, do you want to get us going?

M.L. I have a project in New York City where I build boats and I take people on the water called *Tide and Current Taxi*. The taxi is a rowboat, and we use the tide to navigate New York City. People tell me where they want to go, and then I study tidal charts and decide when we can go based on the tidal currents. I've been doing that project for twenty years. It started out because I thought, "We're using the wrong map of the city to navigate this incredibly well-known traversed space." But the more I have done it, the more I realize it's the real map of the city. It's this rhythm that the landscape is actually governed by. I love thinking about me and my passengers tapping into this different kind of clock. I grew up around the water. My dad was in the military, he loved canoeing and kayaking, and that's what inspired the project.

In the early aughts you didn't see a lot of kayakers and canoers in New York City. And I thought, "Why aren't people using this space way more right now?" It has increased since I've lived there. The project is about the challenges of doing that as well as the discoveries that we make. Sometimes I make a blog about the project. That's how most people encounter it. Although hundreds of people have been with me in the boat, many more people view it online. The project also shows that the way the city is constructed makes it hard to access this incredible natural and communal resource.

B.H. Thank you. Athena, do you want to pick up on where you situate yourself in relationship to the shoreline?

A.L. We grew up around the Anchorage area very much close to shorelines. It's a very different type of shoreline up in Alaska, though. Growing up there embedded a lot of really deep senses of the power of water, the power of the shoreline. We used to go camping in the interior of Alaska every year. There are very deep lakes there that are glacially carved and no shoreline. We would be standing on the edge of the tundra, where the tundra meets these deep, deep lakes that were always filled with mystery because they were so dark. You saw maybe a couple of feet of water along the edge but then it dropped into blackness. We grew up understanding that there is something very much to be respected there because if you were to fall in, you could be dead from hypothermia within minutes. The glacier is moving along through the land they deposited. They moved a lot of silt from underground over into the rivers that empty into the inlets and into the ocean.

Coming from Alaska, my experience on the East Coast was limited and I had to find my own way to connect to the shoreline of the Hudson. Having an art background, I learned about the Hudson River School painters. Pulling that together with my own Lakota and Ojibwe background and growing up among Indigenous communities in Alaska allowed me certain privileges of history. Learning about the Hudson River and the history of the Dutch and the British and the founding of New York City was quite fascinating, especially as a place of trade and atrocities. I'm always looking at the horrors that have happened in a place and the traumas that are rooted there. As I began moving up and down the river and getting to know Brooklyn, getting to know New York Harbor, and just looking at the vast volume of natural material in the water, I saw these as forces of nature in themselves. I was also looking at the power of industry. Now that I live farther upriver in Peekskill, I go down to the shoreline often. I did not know the full history of industry along the Hudson River. There were over twelve foundries around the Peekskill area alone.

I was going down to the shoreline almost daily and was impressed by the massive amount of material embedded in the shoreline. When the tide went out, you could totally see it. It looked like lava. I learned that it was slag from all of the foundries. They remind me of Japanese and Chinese philosopher's stones. It's amazing that you can pull the detritus, the remnants of human industry, out of the river. Human history is transformed by the river eating away at it. I started finding this large, bizarrely fused, molten-like shaped glass; not bottles, more like beach glass, broken and worn. I would bring them back to the studio because I'm fascinated by how forces of water and the heat of the foundries have shaped these materials. There are also natural materials that we pull out as well—driftwood that has fallen from who knows where and has wound up along the shorelines after storms. I'm thinking about these types of things and events.

B.H. Thank you. Courtney, you introduced yourself as being very much involved with water and coming from a nation that has been very close to whaling.

C.L. It may be ironic that about a quarter of my reservation is flooded right now. I've been talking to my family all morning to see how we're going to regroup from not having electricity and having water rise into our homes. The shoreline has been a lot of different things for me as a Shinnecock person growing up. I would probably tie erasure to shoreline and erasure to Eastern Indigenous people. I have been thinking about the shoreline and erasure in the Land Back movement and initiative and that people tend to claim that all of New York is Lenape territory when that's not true. When we think about the Hudson and the water and our trade relationships, I can tie my truth to my knowledge of who I am and where I come from and the science behind it. To be more accurate about our shared histories, our quahog shells are purple and white. They only grow in eastern coastal area communities, and they were traded up to make the Haudenosaunee Wampum Treaty belts and everything thereafter. So, if anybody wanted to actually look a little bit deeper to understand indigeneity, they would ask, "Where did the shell come from?" That shell came from us, and it came from the Wampanoag, and it came from the Pequot, and it came from the Narragansett, and it came from where it came from.

More recently, with the work I've been doing with *Breach,* I also think of shoreline in relationship to Alzheimer's. There's this idea of loss of memory, and there's different ways of having to translate conversations to bring back and build upon memory. As I have experienced that with my grandmother, I have also metaphorically found that relationship to the shoreline. How do we remember the shoreline as it was and the shoreline as it currently is and knowing that the shoreline is going to continue to degrade? I'm often working with clay. I have to keep working while I'm talking, and I'm coiling right now. But I have a lot of knowledge about where clay comes from. It's a blessing that everybody's doing ecological water work to make the Hudson River cleaner. But that's also a conversation that is a hindrance to our Shinnecock community and environmental department when we're trying to write for grants that certain nonprofits are writing for in the Hudson that are often just private, nonprofit, or nonprofit tax write-offs for people. For our community, we need these grants in order to keep what territory we have left. The oysters that you're eating in the city historically would not necessarily be coming from the Hudson River. The oysters filter the water and it's not the best to filter in that territory. So, people grab the oysters from the east and bring them to the restaurants in New York City. Oysters were large and plentiful, and the knowledge of how to live with oysters has just depleted. People don't understand what it means to have natural refuge which then aids in regrowth. Not all waste is bad waste if you have ecological knowledge and relationship to place. All the oysters that are currently being used to build shoreline restoration projects in New York City are coming from our territory. Where is the reciprocity and understanding when people want to talk about decolonization and land back? And where is that conversation when it relates to actually losing their land to water and shoreline?

B.H. Thank you for sharing so many points of contact with topics that we've been thinking about. Jean-Marc, you're looking at different industries and the production of culture that is not in the water but more inland. Do you want to talk a little bit about those relationships?

J.S. Thank you. What really preoccupies me as a public artist is relationships. My interaction with the shoreline started about fourteen years ago in Beacon, where the remnants of brickyards are obvious today. All the detritus was used to extend the shoreline out into the river. Some marine folks will tell you there's a lot that's hidden, one being the giant glacial bed of clay that goes all the way from Albany to Staten Island, which explains the two-hundred-plus brickyards that dotted the shoreline up and down the river in Kingston, Poughkeepsie, Beacon, and Newburgh. Every town had a brickyard at one point, Haverstraw having the biggest concentration of them.

This hidden history is buried now, and it's the thread I keep pulling on to look at the historical record. You realize that the Hudson River School is an ideological representation of the area that is not as accurate as they pretend it to be. It would have had giant smokestacks belching out and all the coal-fired brickmaking plants. It was not a fun place to live on the shoreline in the eighteenth and nineteenth centuries. That's what the historical record lays out. I was describing the hometown of Sojourner, Sojourner Truth being quite significant and being the first Black woman to sue a white man in New York State and win. So, testing the laws of the abolition of slavery in New York State is something I'm aware of when looking at the Hudson River. It's a kind of a tributary of the triangle trade. Hudson Valley farmers farmed the wheat that was sent to the Caribbean in exchange for the rum and the slaves that would have come from the Senegal and Gambia region of Africa after having been traded or been in contact with the early Dutch and then with the arrival of the English—eventually, English colonists. So, when I say there is hidden history in the area, it's like it's just underneath the surface. It's everywhere.

A lot of my work is essentially a kind of ongoing history lesson that I keep sharing with people, some who become aware for the first time that there was slavery in New York. It comes as a surprise—the North being the one who won the Civil War, as if that somehow washes everyone's hands clean. There's a great organization called Forge Project up in the Taconic region whose territory essentially is the place where Robert Livingston, the chancellor and one of the biggest enslavers in the area, managed to convince an unsuspecting colonial governor to sign off on the merging of two properties that he said were right next to each other. It turns out they were about sixty miles apart, and his 2,400 acres suddenly, magically, turned into 165,000. It's this kind of history that makes me appreciate learning about Native histories as well.

C.L. Yeah, I would just ask that when people are learning about places, they also look to understand how these entities came to be. They often also came to be from extreme wealth. The Forge Project became the Forge Project because a very wealthy family supported that initiative. I just always ask people to dig a bit deeper in terms of who is rewriting history.

They own an equestrian farm. They have millions of dollars. I don't know how much, but as someone who is Shinnecock and grew up around extreme wealth and power, I would just ask people to dig a bit deeper in terms of these Land Back initiatives to understand in this welcoming who might we be erasing that was not removed from those territories.

B.H. So we have these layers. We have natural layers, we have marine biology, we have the land mass, we have cultures.

C.L. One thing I appreciate about *Shifting Shorelines* is that you're having many conversations and entry points into the shoreline, and that's what a conversation needs to be. There's no one solid answer to a shoreline or how one defines it.

B.H. Thank you. If you were to physically be somewhere, in the water itself, or on the shoreline, or on a hill, or in an airplane, or in a helicopter, which would be most comfortable to you? Would you be most comfortable standing, sitting, lying down, swimming, or whatever it may be? Why do you feel more comfortable in that position? Do you have a sense why? All of you touched on this a little bit, but I would like to dig deeper into this idea of positionality in relationship to the Hudson specifically. What perspectives do you have because of it?

M.L. Well, of course I love this question. My art project is primarily in the river, looking back at the shoreline. I think about a kind of moving perspective being in the water, looking at this incredibly solid, rigid city. I was listening to Athena talk about those slag pieces and pulling objects out of the muck. I was thinking about the three positions you propose and the question of being on the edge, in the water, or overlooking the water. The work of mine that is going to be in this exhibition is a fourth category of the shoreline—that is, exposing what's under the shoreline that's under the river. Low tide is a really beautiful time to look at the water because you just get a glimpse of what's underneath when the tide recedes. Especially on the Hudson, you can start to see creatures that are exposed by the tide, especially in a negative tide. And then the new moon or the full moon affects things. In the city, you can start to see bulkheads and structures exposed by the low tide. We can never really comprehend what the bottom of the river looks like. Even a scuba diver doesn't see the whole bottom of the river. I love thinking about land that's under the river. We have to hold it in our imagination. And so, I've been thinking about that space beneath the mud and the muck and the invisibility of that space under the water. That's my fourth category.

A.L. Thank you. I always forget about that. I really like what Marie was saying about the shoreline and the currents and the tide. Because for me, it's always fascinating to think about the rushing of the tide or the quiet, the various powers of the tide and what they churn on the floor and on the surface. What is the churning water depositing along the shoreline? All these objects, artifacts, natural, man-made, human made. I recently spent a lot of time at the shoreline looking specifically at high and low tides. With the recent flood that we had, I was eager to get down to the shoreline to see what came up. I was watching these incredible shapes at the edge. As much as I want to be in the water, and go in deep, I just don't, I'm not equipped for that. But I am equipped for treading the shore on the edge there. And that, for me, continues to be a source of mystery. That kind of power and movement of material.

B.H. Thank you. Courtney, do you want to add any more about the shoreline?

C.L. Overlooking. I go to the water to heal, and I also go to listen. And if you listen, you know what it's telling you. And sometimes it's pretty obvious. And sometimes you just have to stay a little while, that's all.

B.H. And Jean-Marc?

J.M. I couldn't really find a position other than askance. I'm looking at representations of the Hudson River shoreline with a degree of skepticism and looking at completing the absences in those histories. Sometimes it is harvesting detritus from the brickyards, just trying to see the unseen, even though there's no evidence, it doesn't necessarily announce itself.

B.H. Thank you. I think everyone here is in touch with this idea of things that are there that maybe are not visible at all or not readily visible. I'm sorry to hear your community was flooded, Courtney, and I wish everyone the best of luck in tackling the water challenges from the bay.

Catalogue of the Exhibition

All works mentioned in this section appear in the exhibition. Works that are illustrated in this volume are indicated by a figure number.

ART

1. American (18th century)
Philipse Manor Hall, 1784
Ink on paper, Sight: 19¼ × 26¼ in. (48.9 × 66.7 cm)
Historic Hudson Valley, Pocantico Hills, New York, Gift of La Duchesse de Talleyrand (PM.65.866)
Fig. 6

2. Henry Ary (1807–1859)
View of the Hudson, 1852
Oil on canvas, 26 × 36 in. (66.0 × 91.4 cm)
Albany Institute of History & Art, Purchase (1977.17)
Fig. 35

3. Victor Gifford Audubon (1809–1860)
View of the Hudson River, ca. 1845
Oil on canvas, 48 × 72 in. (121.9 × 182.9 cm)
Museum of the City of New York, Gift of Miss Alice Lawrence (1938.188)
Fig. 7

4. Alvin Baltrop (1948–2004)
The Piers (to confirm title), 1975–86
Gelatin silver print, 8 × 9¾ in. (20.3 × 24.75 cm)
Estate of Alvin Baltrop, courtesy of Galerie Buchholz, New York

5. Alvin Baltrop (1948–2004)
The Piers (collapsed architecture, couple buttfucking), 1975–86
Gelatin silver print, 4¾ × 7 in. (21.1 × 17.75 cm)
Estate of Alvin Baltrop, courtesy of Galerie Buchholz, New York

6. Alvin Baltrop (1948–2004)
The Piers (three figures in warehouse), 1975–86
Gelatin silver print, 3⅞ × 5⅞ in. (10 × 15 cm)
Estate of Alvin Baltrop, courtesy of Galerie Buchholz, New York

7. Alvin Baltrop (1948–2004)
The Piers (two people lying by Hudson River), 1975–86
Gelatin silver print, 4½ × 7 in. (11.4 × 17.8 cm)
Estate of Alvin Baltrop, courtesy of Galerie Buchholz, New York

8. Alvin Baltrop (1948–2004)
The Piers (figure on dock), 1975–86
Gelatin silver print, 6 × 2½ in. (15.2 × 6.4 cm)
Estate of Alvin Baltrop, courtesy of Galerie Buchholz, New York

9. Alvin Baltrop (1948–2004)
The Piers (dock at night), 1975–86
Gelatin silver print, 4½ × 6¾ in. (11.4 × 17.1 cm)
Estate of Alvin Baltrop, courtesy of Galerie Buchholz, New York

10. Gifford Reynolds Beal (1879–1956)
Freight Yards, 1915
Oil on canvas, 36 × 48 in. (91.4 × 121.9 cm)
Everson Museum of Art, Syracuse, Museum purchase with funds from American Art Fund (PC 15.80)
Fig. 42

11. Julie Hart Beers (1835–1913)
Cows in a Landscape, 1869
Oil on canvas, 9⅛ × 14⁹⁄₁₆ in. (23.2 × 37.0 cm)
Lebanon Valley College Fine Art Collection, Annville, Pennsylvania (2020.1.45)
Fig. 17

12. George Bellows (1882–1925)
Rain on the River, 1908
Oil on canvas, 32⅜ × 38¼ in. (82.2 × 97.2 cm)
RISD Museum, Providence, Rhode Island, Jesse Metcalf Fund (15.063)
Fig. 20

13. Daniel Putnam Brinley (1879–1963)
Hudson River View (Sugar Factory at Yonkers), ca. 1915
Oil on canvas, 30⅛ × 31⅞ in. (76.5 × 80.9 cm)
Hudson River Museum, Yonkers, Museum Purchase, 1995 (95.31)
Fig. 22

14. Johann Hermann Carmiencke (1810–1864)
Poughkeepsie Iron Works (Bech's Furnace), 1856
Oil on canvas, 29 × 36¼ in. (73.7 × 92.1 cm)
Yale University Art Gallery, New Haven, Bequest of Evelyn A. Cummins (1971.111.5)
Fig. 36

15. Center for Land Use Interpretation
A Journey up the Hudson River from the Battery to Troy from the series *Up River: Points of Interest on the Hudson*, 2006
Digital photographs
The Center for Land Use Interpretation, Culver City, California
Fig. 48

16. Frederic Edwin Church (1826–1900)
Hudson River with Factory by Moonlight, 1844–45
Brush and oil on paperboard, 8¹¹⁄₁₆ × 8¹⁵⁄₁₆ in. (22.1 × 22.7 cm)
Cooper Hewitt, Smithsonian Design Museum, Smithsonian Institution, Gift of Louis P. Church (1917-4-44)
Fig. 9

17. Thomas Cole (1801–1848)
North Mountain and Catskill Creek, 1838
Oil on canvas, 26⁷⁄₁₆ × 36⁷⁄₁₆ in. (67.2 × 92.6 cm)
Yale University Art Gallery, New Haven, Gift of Anne Osborn Prentice (1981.56)
Fig. 15

18. Glen O. Coleman (1887–1932)
Fort Lee Ferry, 1923
Oil on canvas, 25 × 29⅞ in. (63.5 × 75.9 cm)
Brooklyn Museum, John B. Woodward Memorial Fund (30.1153)

19. Samuel Colman (1832–1920)
Quarry Works, Hastings on Hudson, before December 1875
Watercolor on paper, 7¹¹⁄₁₆ × 15⁷⁄₁₆ in. (19.5 × 39.2 cm)
Museum of Fine Arts, Boston, Gift of Maxim Karolik for the M. and M. Karolik Collection of American Watercolors and Drawings, 1800–1875 (61.339)
Fig. 40

20. John V. Cornell (1813–1849)
Steamboat "Iron Witch," 1846
Oil on canvas, 26 × 38 in. (66.0 × 96.5 cm)
New-York Historical Society, Gift of Samuel V. Hoffman (1929.146)

21. Jasper F. Cropsey (1823–1900)
The Hudson River at Hastings, 1889
Oil on canvas, 8⅝ × 20 in. (21.9 × 50.8 cm)
Newington-Cropsey Foundation (168)
Fig. 43

22. Henry Golden Dearth (1864–1918)
Ice Boats on the Hudson, ca. 1888–98
Oil on canvas, 18 × 29⅛ in. (45.7 × 74.0 cm)
Telfair Museums, Savannah, Georgia, Gift of Gari Melchers (1908.6)
Fig. 1

23. Aaron Douglas (1899–1979)
Inwood Power Plant, 1934
Oil on canvas, 20 × 22 in. (50.8 × 55.8 cm)
Collection of the Hampton University Museum, Hampton, Virginia (67.112)
Fig. 24

24. Joellyn Duesberry (1944–2016)
Cement Factory, Hudson River, 1983–84
Oil on linen, 20 × 36 in (50.8 × 91.4 cm)
Hudson River Museum, Yonkers, Gift of the artist, 2007 (2007.01.01)
Fig. 41

25. Ernest Fiene (1894–1965)
Hudson River Boat, 1928
Lithograph, 12⁵⁄₁₆ × 15⅜ in. (31.3 × 39.1 cm)
The Museum of Modern Art, New York, Gift of Abby Aldrich Rockefeller, 1940 (848.1940)

26. Kryn Frederycks (17th century)
T'Fort Nieuw Amsterdam op de Manhatans, 1651
Engraving, printed in *Beschrijvinge van Virginia, Nieuw Nederlandt, Nieuw Engelandt* (1651), 3¼ × 4⅞ in. (8.3 × 12.4 cm)
Isaac Newton Phelps Stokes Collection, Miriam and Ira D. Wallach Division of Art, Prints and Photographs, The New York Public Library, Astor, Lenox, and Tilden Foundations (188138)
Fig. 3

27. Reva Fuhrman (b. 1985)
Beaded dance regalia: hat, yoke, cuffs and belt, 2018
Velvet and cotton fabric with glass beads
Hat: 3¼ × 7½ in. (8.3 × 19.1 cm); yoke: 27½ × 15½ in. (68.9 × 39.4 cm); cuffs: 7¾ × 9¾ in. (19.7 x 24.8 cm); belt: 10½ × 34 in. (26.7 × 86.4 cm)
New York State Museum, Albany (E-2018.39.1–4)
Fig. 5

28. Emil Ganso (1895–1941)
New York City Nocturne, 1929
Oil on canvas, 20 × 16 in (50.8 × 40.6 cm)
D. Wigmore Fine Art, Inc., New York

29. Marie-François-Régis Gignoux (1814–1882)
Haverstraw on the Hudson, 1860–65
Oil on canvas, 13½ × 35¼ in. (33.0 × 88.9 cm)
Albany Institute of History & Art, Purchase (1951.68)
Fig. 18

30. Shi Guorui (b. 1964)
View of Catskill Mountains, New York, February 6–7, 2019
Gelatin silver print, 45 × 115 in. (114.3 × 292.1 cm)
Collection of the artist
Fig. 26

31. David Hammons (b. 1943)
Day's End, 2014–21
Four photographs of site-specific installation, Hudson River Park, by Jason Schmidt, 20 × 24 in. (50.8 × 61.0 cm)
Courtesy Whitney Museum of American Art, New York / Licensed by Scala / Art Resource, NY
Fig. 49

32. Palmer Hayden (1890–1973)
South Ferry, ca. 1930
Oil on canvas, 24 × 32 in. (61.0 × 81.3 cm)
The Johnson Collection, Spartanburg, South Carolina
Fig. 47

33. Donna Hogerhuis (b. 1958)
Lidded Basket, 2024
Machine-made ash splints, 10½ × 13½ × 8½ in. (26.7 × 34.3 × 21.6 cm)
Collection of the artist

34. Edward Hopper (1882–1967)
Hook Mountain, Nyack, ca. 1899
Watercolor on paper, 5 1/16 × 6⅞ in. (12.9 × 17.5 cm)
Whitney Museum of American Art, New York, Josephine N. Hopper Bequest (70.1558.55)
Fig. 37

35. Every Ocean Hughes (b. 1977)
The Piers Untitled (#4), 2009–23
Gelatin silver print, custom frame, 33 × 33 in. (83.8 × 83.9 cm)
Collection of the artist

36. Every Ocean Hughes (b. 1977)
The Piers Untitled (#5), 2009–23
Gelatin silver print, custom frame, 33 × 33 in. (83.8 × 83.9 cm)
Collection of the artist
Fig. 38

37. Every Ocean Hughes (b. 1977)
The Piers Untitled (#7), 2009–23
Gelatin silver print, custom frame, 33 × 33 in. (83.8 × 83.9 cm)
Collection of the artist
Fig. 39

38. Every Ocean Hughes (b. 1977)
The Piers Untitled (#13), 2009–23
Gelatin silver print, custom frame, 33 × 33 in. (83.8 × 83.9 cm)
Collection of the artist

39. William Henry Jackson (1843–1942)
Spuyten Duyvil, First View of the Hudson, ca. 1890
Albumen print mounted on mat board, 17¼ × 21⅛ in. (43.9 × 53.6 cm)
Prints and Photographs Division, Library of Congress, Washington, D.C. (Lot 12697, no. 1)

40. William Henry Jackson (1843–1942)
Palisades of the Hudson River, N.Y., ca. 1900
Photomechanical print: Photochrom, color, 7 15/16 × 23 9/16 in. (20.2 × 59.8 cm)
Prints and Photographs Division, Library of Congress, Washington, D.C. (Lot 3886, no. 13)

41. Yvonne Jacquette (1934–2023)
Hudson River Diptych, 2007
Woodcut, 21 × 40 in. (53.3 × 101.6 cm)
Estate of Yvonne Jacquette, courtesy of Mary Ryan Gallery, New York
Fig. 52

42. David Johnson (1827–1908)
View from Garrison, West Point, New York, 1870
Oil on canvas, 18¼ × 30 in. (46.4 × 76.2 cm)
Dallas Museum of Art, The Patsy Lacy Griffith Collection, Gift of Patsy Lacy Griffith by exchange, and General Acquisitions Fund (2012.6)

43. Abraham Leon Kroll (1884–1974)
View of Manhattan from the Terminal Yards, Weehawken, New Jersey, 1913
Oil on canvas, 36 × 48 in. (91.4 × 121.9 cm)
Montgomery Museum of Fine Arts, Alabama, Association Purchase: Art Acquisitions Fund and Gifts of Mr. E. Baldwin Goetter in memory of his parents, Mr. and Mrs. Joseph Goetter and his sister, Mrs. Mabel Goetter Godchaux, George H. Todd, Dr. Sonia Lupian, Mrs. James Norment Baker in memory of Dr. James Nornent Baker, Gift in memory of Mrs. Onita Henderson sponsored by Mrs. Nash Read, Mrs. J. S. Hough, Bessic D. McGavock, and Virginia Barnes, by exchange (2000.1)
Fig. 21

44. Athena LaTocha (b. 1969)
The Discovery of Slowness, 2022
Shellac ink, silt, mica on paper, lead, steel, 46 × 122 in. (116.8 × 309.8 cm)
Antonio Murzi and Diana Morgan Collection
Fig. 56

45. Ernest Lawson (1873–1939)
Spuyten Duyvil Creek, ca. 1914
Oil on linen, 25 × 30 in. (63.5 × 76.2 cm)
Wichita Art Museum, Bequest of Glenn L. and Jayne Seydell Milburn (2017.20)
Fig. 25

46. An-My Lê (b. 1960)
Truck Load Out (River), *Hudson River, Mill Silos III (Backhoe)*, *Beach Parts*, *Excavator*, and *Fixed Stackers*, from *Trap Rock*, 2006–7
Pigment prints, each 30 × 42 in. (76.2 × 106.7 cm)
Collection of the artist, courtesy of Marian Goodman Gallery, New York
Fig. 62

47. Courtney M. Leonard (b. 1980)
Work in progress at time of publication

48. Marie Lorenz (b. 1973)
Jet Ski, 2024
Steel, gypsum cement, 60 × 54 × 123 in. (152.4 × 137.2 × 312.4 cm)
Collection of the artist
Fig. 65

49. George Benjamin Luks (1867–1933)
Roundhouse at High Bridge, 1909–10
Oil on canvas, 36½ × 32¾ in. (92.7 × 83.2 cm)
Munson-Williams-Proctor-Arts-Institute, Utica, New York, Museum Purchase (50.17)
Fig. 23

50. John Marin (1870–1953)
Lower Manhattan from River, No. 1, 1921
Watercolor, charcoal, graphite, on paper, 21⅞ × 26½ in. (55.6 × 67.3 cm)
The Metropolitan Museum of Art, New York, Alfred Stieglitz Collection, 1949 (49.70.122)
Fig. 27

51. Reginald Marsh (1898–1954)
Tugboat and New York City Skyline, 1933
Watercolor on paper, 14 × 20 in. (35.6 × 50.8 cm)
D. Wigmore Fine Art, Inc., New York

52. Gordon Matta-Clark (1943–1978)
Day's End, 1975
Super 8mm film on HD video, 23 min. 10 sec.
Electronic Arts Intermix

53. Alex Matthew (19th century)
Oystering at Prince's Bay, ca. 1853
Oil on canvas, 25⅛ × 33¼ in. (63.8 × 84.5 cm)
Staten Island Historical Society, Collection of Historic Richmond Town (P01.0057)
Fig. 12

54. Alan Michelson (b. 1953)
Shattemuc, 2009
HD video, stereo soundtrack with original music by Laura Ortman, 31 min.
Collection of the artist
Fig. 2

55. Charles Frederick William Mielatz (1860–1919)
Hudson River, Oyster Market Near Christopher Street, 1898
Lithograph, Sheet: 7¾ × 10⅛ in. (19.7 × 25.7 cm)
Museum of the City of New York, Arthur H. Scribner Collection, 1941 (41.62.20)
Fig. 46

56. After Jacques Gerard Milbert (1766–1840)
Haverstraw or Warren Landing, 1828–29
Lithograph, 7¹¹⁄₁₆ × 11³⁄₁₆ in. (19.5 × 28.4 cm)
Yale University Art Gallery, New Haven, Mabel Brady Garvan Collection (1946.9.1947)
Fig. 13

57. Thomas Moran (1837–1926)
Lower Manhattan from Communipaw, New Jersey, 1880
Oil on canvas, 25¼ × 45¼ in. (64.1 × 114.9 cm)
Washington County Museum of Fine Arts, Hagerstown, Maryland (A303.41.01)
Fig. 19

58. William H. Moschett (19th century)
Oyster Sloop in the Kill van Kull, ca. 1850s–1860s
Oil on canvas, 18½ × 24 in. (47.0 × 61.0 cm)
Staten Island Historical Society, Collection of Historic Richmond Town (P01.0214)

59. Ruth Orkin (1921–1985)
Boy Jumping into Hudson River, NYC, 1948
Gelatin silver print, 14 × 10$^{15}/_{16}$ in. (35.6 × 27.8 cm)
The Jewish Museum, New York, Purchase: Horace W. Goldsmith Foundation Fund (2008.98)
Fig. 50

60. Anthony Papa (b. 1960)
View of the Hudson River from Sing Sing #3, 1992
Mixed media on 100% rag paper, 15 × 22 in. (38.1 × 55.9 cm)
Collection of the artist

61. Anthony Papa (b. 1960)
View of the Hudson River from Sing Sing 7 Building Yard, 1992
Mixed media on 100% rag paper, 15 × 22 in. (38.1 × 55.9 cm)
Collection of the artist
Fig. 45

62. Anthony Papa (b. 1960)
View of the Hudson River from Sing Sing Overview 1, 1994
Mixed media on 100% rag paper, 15 × 22 in. (38.1 × 55.9 cm)
Collection of the artist

63. Anthony Papa (b. 1960)
White Butterflies, Blue Hudson, 1995
Acrylic and oil on canvas board, 22 × 24 in. (55.9 × 61.0 cm)
Collection of the artist
Fig. 44

64. Anthony Papa (b. 1960)
View of the Hudson River with Guard Tower, 1995
Acrylic on canvas board, 20 × 24 in. (50.8 × 61.0 cm)
Collection of the artist

65. Lisa Sanditz (b. 1973)
Tivoli Bay, 2016
Oil on canvas, 54 × 70 in. (137.2 × 177.8 cm)
Collection of the artist, courtesy Huxley Parlour Gallery, London
Fig. 33

66. Henry Schnakenberg (1892–1970)
Edgewater, NJ, 1938
Oil on canvas, 22 × 32 in. (55.9 × 81.3 cm)
Weatherspoon Art Museum, University of North Carolina Greensboro, Museum purchase with funds from the Laura Weill Cone Acquisition Endowment, 2014 (2014.20)

67. Jean-Marc Superville Sovak (b. 1976)
a-Historical Landscape: Hudson Highlands/Gospel of Slavery, 2019
Monoprint on a found engraving, 8 × 10 in. (20.3 × 25.4 cm)
Collection of the artist

68. Jean-Marc Superville Sovak (b. 1976)
a-Historical Landscape: Albany/On to Liberty, 2019
Monoprint on found engraving, 8 × 10 in. (20.3 × 25.4 cm)
Collection of the artist

69. Jean-Marc Superville Sovak (b. 1976)
a-Historical Landscape: Cold Spring/Fugitive Slave Act, 2019
Monoprint on found engraving, 8 × 10 in. (20.3 × 25.4 cm)
Collection of the artist

70. Jean-Marc Superville Sovak (b. 1976)
a-Historical Landscape: Peekskill/Runaways, 2019
Monoprint on found engraving, 8 × 10 in. (20.3 × 25.4 cm)
Collection of the artist
Fig. 60

71. Jean-Marc Superville Sovak (b. 1976)
a-Historical Landscape: New York from Weehawken/Rose Butler, 2019
Monoprint on found engraving, 8 × 10 in. (20.3 × 25.4 cm)
Collection of the artist

72. Jean-Marc Superville Sovak (b. 1976)
a-Historical Landscape: Hyde Park, 2019
Monoprint on a found engraving, 8 × 10 in. (20.3 × 25.4 cm)
Collection of the artist

73. Jean-Marc Superville Sovak (b. 1976)
Freeborn Trails (North Star), 2021; re-creation, 2024
Bricks, pebbles, diam. 8 ft.
Collection of the artist
Fig. 59

74. Alfred Stieglitz (1864–1946)
The City Across the River: From the Journal "Camera Work," 1911
Photogravure, 7⅞ × 6⁵⁄₁₆ in. (20.0 × 16.19 cm)
Philadelphia Museum of Art, Gift of Carl Zigrosser, 1973 (1973-87-42)

75. Joseph Vollmering (1810–1887)
View on the Hudson Near Sing Sing, New York, 1845–50
Oil on canvas, 21½ × 28 in. (54.6 × 71.1 cm)
New-York Historical Society, Purchase, Watson Fund (1978.57)
Fig. 8

76. John Ferguson Weir (1841–1926)
View of the Highlands from West Point, 1862
Oil on linen, 20 × 34 in (50.8 × 86.4 cm)
New-York Historical Society, The Robert L. Stuart Collection, Gift of his widow Mrs. Mary Stuart (S-224)

77. John Ferguson Weir (1841–1926)
The Gun Foundry, 1866
Oil on canvas, 65 × 80 in. (165.1 × 203.2 cm)
Putnam History Museum, Cold Spring, New York
Fig. 10

78. Worthington Whittredge (1820–1910)
Shad Fishing on the Hudson, ca. 1875
Oil on canvas, 11½ × 13½ in. (29.2 × 34.3 cm)
Philadelphia Museum of Art, Gift of Marguerite and Gerry Lenfest, 2008 (2008.124.3)
Fig. 16

MATERIAL CULTURE, ARCHAEOLOGICAL ARTIFACTS, SCIENTIFIC MATERIAL

79. Algonquian, Mohican
Plaited basket, n.d.
Split ash and ink, 10½ × 16 × 13 in. (26.7 × 40.6 × 33 cm)
New York State Museum, Albany (E-39529A-B)
Fig. 30

80. Algonquian, Mohican
Pair of leggings, ca. 1800–1820
Green woolen trade fabric embellished with green trim and white beadwork, 16½ × 8 in. (41.9 × 20.3 cm)
New York State Museum, Albany (E-36296A-B)

81. Algonquian, Mohican
Moccasins, ca. 1800–1820
Dearskin embellished with ribbon and white beadwork, 9 × 4 in. (22.9 × 10.2 cm)
New York State Museum, Albany (E-36295A-B)
Fig. 4

82. Unmarked oyster jar, 18th century
Ceramic with salt glaze, 7 × 4¾ in. (17.8 × 12.1 cm)
Collection of Chris Pickerell

83. Unmarked oyster jar, 18th century
Ceramic with salt glaze, 7 × 4¾ in. (17.8 × 12.1 cm)
Collection of Chris Pickerell

84. Thomas W. Commeraw (ca. 1772–1823)
Oyster jar marked "Daniel Johnson and Co.," 1799–1804
Ceramic with salt glaze, 9 × 5¾ in. (22.9 × 14.2 cm)
Collection of Chris Pickerell
Fig. 11 (center)

85. Thomas W. Commeraw (ca. 1772–1823)
Oyster jar marked "Daniel Johnson and Co.," 1799–1804
Ceramic with salt glaze, 5¼ × 3½ in. (13.3 × 8.9 cm)
Collection of Chris Pickerell
Fig. 11 (left)

86. Thomas W. Commeraw (ca. 1772–1823)
Oyster jar marked "Daniel Johnson and Co.," 1799–1804
Ceramic with salt glaze, 5¾ × 3½ in. (14.6 × 8.9 cm)
Collection of Chris Pickerell

87. Oyster jar marked "Henry Scott," 1820–40
Ceramic with Albany slip glaze, 6 × 3¾ in. (15.2 × 9.5 cm)
Collection of Chris Pickerell
Fig. 11 (right)

88. Unmarked oyster jar, presumably from Norwalk, Connecticut, 1820–40
Ceramic with salt glaze, 8 × 4¼ in. (20.3 × 10.8 cm)
Collection of Chris Pickerell

89. Unmarked oyster jar, presumably from Norwalk, Connecticut, 1820–40
Ceramic with salt glaze, 9¼ × 6 in. (23.5 × 15.2 cm)
Collection of Chris Pickerell

90. Oyster or fish jar marked "F. H. Johnson and Co.," early 20th century
Ceramic with salt glaze, 5¾ × 5 in. (14.6 × 12.7 cm)
Collection of Chris Pickerell

91. Unmarked oyster or fish jar with handle, early 20th century
Ceramic with salt glaze, 7½ × 6 in. (19.1 × 15.2 cm)
Collection of Chris Pickerell

92. Eastern Oyster (Crassostrea virginica) shells excavated from a domestic trash midden associated with New York City's first Almshouse (1736–97) located in what is now New York City Hall Park, ca. 1736–97
NYC Archaeological Repository: The Nan A. Rothschild Research Center

93. Mohican, 5500 BCE–1609 CE
Cache of five pebble net sinkers from the east bank of the Mahicannituck (Hudson River) at Newton Hook
Collection of Stockbridge-Munsee Band of Mohicans
Fig. 31

94. Mohican/Munsee/Lenape/Woodland Culture/Indigenous, 5500 BCE–1609 CE
Cache of pebble net sinkers
Collection of Stockbridge-Munsee Band of Mohicans

95. Male Atlantic sturgeon (*Acipenser oxyrinchus*) scutes from a dying fish caught during a research initiative June 1, 2022, at Mahicannituck (Hudson River) near Hyde Park, New York
Collection of Stockbridge-Munsee Band of Mohicans
Fig. 32

96. Piermont Marsh Sediment Core
One-meter sample taken in 2022 showing a four-hundred-year history
Collection of Dorothy M. Peteet

ABOUT THE CONTRIBUTORS

Ross Barrett is a historian of nineteenth- and twentieth-century American art and visual culture. He is the author of *Rendering Violence: Riots, Strikes, and Upheaval in Nineteenth-Century American Art* (2014) and *Speculative Landscapes: American Art and Real Estate in the Nineteenth Century* (2022) and co-editor, with Daniel Worden, of *Oil Culture* (2014). He is working on a new book that will explore the ways that American artists and everyday creators reckoned with the experience and memory of Atlantic hurricanes during the nineteenth century.

Annette Blaugrund, former director of the National Academy Museum and School of Fine Arts (1997–2007), has published and lectured widely on American art and culture. She was the Andrew W. Mellon senior curator at the New-York Historical Society and a curator at the Brooklyn Museum and the Pennsylvania Academy of the Fine Arts. She has written sixteen books about American art, including *Paris 1889: American Artists at the Universal Exposition* (1989), *The Tenth Street Studio Building* (1997), *John James Audubon* (1999), and *Thomas Cole: The Artist as Architect* (2016). In 1992 she was named a Chevalier of the Order of Arts and Letters by the French government and in 2008 received a Lifetime Achievement Award from the National Academy of Design. She has a Ph.D. in art history from Columbia University (1987), where she taught American art (1996–2001). She is consulting curator at the Thomas Cole National Historic Site in Catskill and a member of the Advisory Council of Columbia's Department of Art History.

Tracy Brown became President and Hudson Riverkeeper in 2021. A recognized leader in clean-water advocacy, she brings a multidisciplinary approach to the organization, prioritizing data-driven, community-oriented strategies to realize its mission. During her prior experience at Riverkeeper (2007–14), she was instrumental in developing water-quality monitoring programs and was an architect of New York's Sewage Pollution Right to Know Law. She served as Regional Director of Water Protection at Save the Sound (2014–21), establishing their New York office, which delivers science-based projects that protect the Long Island Sound and increase community resilience. She is a founder of the Peabody Preserve Outdoor Classroom, a nature preserve for hands-on outdoor education for the Tarrytown-Sleepy Hollow public schools students.

Betti-Sue Hertz became Director and Chief Curator at Columbia University's Wallach Art Gallery in 2019. Her curatorial and scholarly work focuses on the intersection of critical visual culture, transnational exchange, and socially relevant issues. Hertz was Director, Longwood Arts Project, Bronx (1992–98); Curator of Contemporary Art, San Diego Museum of Art (2000–2008); Director of Visual Arts, Yerba Buena Center for the Arts (2008–15); and Public Arts Consultant, TLS Landscape Architecture (2015–20). Hertz has organized more than eighty exhibitions during her career. She was a member of Stanford Art Institute's Creative Cities Working Group (2016–19) and a founding member of RepoHistory (1989–2000). Hertz received a B.A. from Goddard College, an M.F.A. from Hunter College, and studied in the Ph.D. Program in Art History at the City University of New York. She has taught social art history and theory at Stanford University, San Francisco Art Institute, and University of California, Berkeley.

Victoria Horrocks is a curator and art historian whose research includes modern and contemporary American art history, the intersection of art and literature, and the construction of historical and personal narrative. She received a B.A. in English from Cornell University, an M.A. in Modern and Contemporary Art from Columbia University, and a Master of Studies in History of Art and Visual Culture from the University of Oxford. Her recent publications include *Photography at the Border: Presence and Loss in Border Cantos and Lost Children Archive* (2023) and articles in *MODA Critical Review*, *Online Gallery*, and *Bowdoin Journal of Art*.

Elizabeth Hutchinson is Tow Associate Professor of Art History at Barnard College, Columbia University. Her research is centered on the relationship between the visual culture of various North American groups and its viewers. Key issues motivating her work include visuality and modernity, transculturation in the arts of the Americas, and comparative analyses of the visual culture of the United States and other colonial cultures. She has written extensively on how Native Americans used "modern" art to negotiate a place for themselves within industrial culture at the turn of the twentieth century. She is the author of *The Indian Craze: Primitivism, Modernism and Transculturation in Native American Art, 1890–1915* (2009). She has received support for her work from the National Endowment for the Humanities, Georgia O'Keeffe Museum, Sterling and Francine Clark Art Institute, and Winterthur.

Athena LaTocha creates massive works on paper exploring the relationship between human-made and natural worlds. The artist has incorporated materials such as ink, lead, earth, and burned wood while responding to the storied and, at times, traumatic histories that are rooted in place. Her work is in the collections of the Virginia Museum of Fine Arts, Richmond; Hessel Museum of Art, Annandale-on-Hudson; Dallas Museum of Art; and Plains Art Museum, Fargo. LaTocha is the recipient of support from numerous organizations, including Robert Rauschenberg Foundation (2013), Wave Hill (2018), Joan Mitchell Foundation (2016, 2019), National Academy Affiliated Fellowship at the American Academy in Rome (2021), Rockefeller Brothers Fund Pocantico Art Prize in Visual Arts (2022), Anonymous Was a Woman Award (2023), and Foundation for Contemporary Arts Visual Arts Grant (2024).

Courtney M. Leonard is an artist and filmmaker involved with the Offshore Art movement. Leonard's work embodies the multiple definitions of "breach," an exploration and documentation of historical ties to water, whales, and material sustainability. In collaboration with national and international museums, cultural institutions, and Indigenous communities in North America, New Zealand, Nova Scotia, and the United States Embassies, Leonard's practice investigates narratives of cultural viability as a reflection of environmental record. Her work is in the permanent collections of the Art in Embassies Program; Crocker Art Museum, Sacramento; Heard Museum, Phoenix; Ceramics Research Center & Archive, Arizona State Art Museum, Tempe; Peabody Essex Museum, Salem, Mass.; Newman Museum of Contemporary Art, Overland Park, Kans.; Museum of the North, University of Alaska, Fairbanks; Mystic Seaport Museum, Stonington, Conn.; and Pomona Museum of Art.

Marie Lorenz is a visual artist living in New York City. In her ongoing project *The Tide and Current Taxi* Lorenz takes participants through New York waterways in boats that she designs and builds, using tidal current to propel them. Recent solo exhibitions include *Waterways* at Susanne Lemberg Usdan Gallery, Bennington College; *Ash Heap/Landfill* at Lupin Foundation Gallery, Lamar Dodd School of Art in Athens, Ga.; and *Ezekia* at Albright-Knox Art Gallery, Buffalo. Her work was included *The Sorcerer's Burden: Contemporary Art and the Anthropological Turn* at The Contemporary, Austin, and *The Commuter Biennial* produced with a grant from the Knight Foundation in Miami. Lorenz recently received Creative Capital and National Endowment for the Arts grants for an opera set and performed along the Newtown Creek in New York City. Other residencies and honors include the Joseph H. Hazen Rome Prize for the American Academy in Rome (2008) and a Harpo Foundation grant for her exhibition at Locust Projects in Miami (2011). Lorenz received a B.F.A. from Rhode Island School of Design and an M.F.A. from Yale and is represented by Jack Hanley Gallery in New York.

May Joseph is the founder of Harmattan Theater, a Professor of Social Science at Pratt Institute, and author of *Nomadic Identities: The Performance of Citizenship* (1999), *Fluid New York: Cosmopolitan Urbanism and the Green Imagination* (2013), *Sea Log: Malabar to New York* (2019), and *Ghosts of Lumumba* (2020). She is a co-author of *Aquatopia: Climate Interventions* (2023) and co-editor of *Terra Aqua: The Amphibious Lifeworlds of Coastal and Maritime South Asia, Performing Hybridity, Critical Climate Studies, Ocean and Island Studies* and *Kaleidoscope: Ethnography, Art, Architecture and Archaeology*. Joseph creates site-specific performances along Dutch and Portuguese maritime routes, exploring climate issues.

Anthony Papa is an artist, writer, and advocate against the war on drugs. His art has been exhibited widely, including at the Whitney Museum, and he has been interviewed on talk shows such as CNN's "Your Money" and in the *New York Times*. Papa served twelve years of a fifteen-to-life sentence for a nonviolent drug crime in New York State. In 1996 he received clemency from Governor George Pataki. In 2016 he received a pardon from Governor Andrew Cuomo and became the first person in New York State history to receive both clemency and a pardon. Mr. Papa is a frequent public speaker and college lecturer on his art and on criminal justice issues. He is the author of *15 to Life: How I Painted My Way to Freedom* (2004) and *This Side of Freedom: Life After Clemency* (2016).

Dorothy M. Peteet is a Senior Research Scientist at NASA/Goddard Institute for Space Studies and Adjunct Professor, Columbia University. She directs the Paleoecology Division of the New Core Lab at Lamont Doherty Earth Observatory of Columbia and in collaboration with GISS climate modelers is studying the Late Pleistocene and Holocene archives of lakes and wetlands. Peteet, a long-time Hudson Valley resident, is interested in how the environment has shifted. Her recent research focuses on human impact documented by forest decline, pollution history, invasive species, and carbon storage in Hudson River marshes.

Christopher Pickerell is a biologist and independent ceramic researcher born and raised on Long Island. He is the Director of the Marine Program for Cornell Cooperative Extension of Suffolk County overseeing a team of scientists and professionals working throughout the New York marine district and surrounding region. A longtime antiques and ceramics enthusiast, he became interested in oyster jars in 2015, when he acquired his first two examples. Dissatisfied with the dearth of literature on the subject, he has spent the last nine years combing original source materials and consulting with archaeologists, historians, and potters to better understand the role of these pots in early New York history. He has a collection of more than forty examples that he uses for research and educational purposes.

Steven Schimmrich is Professor of Geology & Earth Sciences and STEM Department Chair at SUNY Ulster County Community College in Stone Ridge, where he has taught for more than twenty-five years. In his spare time, he is an avid local hiker. His research interests include the geologic history of the Northeast and the interrelationships between geology and the historical development of the Hudson Valley. He is interested in promoting science education and literacy, teaches a yearly summer field course on the geology of the Hudson Valley, and is the author of *Geology of the Hudson Valley: A Billion Years of History* (2020).

Jean-Marc Superville Sovak is an artist whose work "critically fabulates" around silent histories of multiracial identities that make up the DNA of this country. His *a-Historical Landscapes* are nineteenth-century engravings to which he adds images from anti-slavery publications. His public artworks include monuments to Afro-Dutch pioneers in Rockland County and a memorial to some of the earliest Africans to arrive in Rhode Island. He received an M.F.A. from Bard College and an Individual Artist's Commission and an Empowering Artist Award from Art Mid-Hudson. His work has been exhibited at Aldrich Contemporary Art Museum, Ridgefield, Conn.; Arts Westchester, White Plains; Socrates Sculpture Park, Queens; and Recess Arts, Brooklyn; and is in the permanent collections of the Frances Lehman Loeb Art Center, Vassar College, and the Dorsky Museum of Art, New Paltz. Jean-Marc has been a visiting artist at Bard College, SUNY New Paltz, Columbia University, and Vassar College.

Jonathan Weinberg is Curator of the Maurice Sendak Foundation and author of several books, including *Ambition and Love in Modern American Art* (2001) and *Pier Groups: Art and Sex Along the New York Waterfront* (2019). He was lead curator of the exhibitions *Art After Stonewall, 1969–89* (2019) and is curator of *Wild Things Are Happening: The Art of Maurice Sendak* (2024). In 2022 his GENESIS paintings and prints were on view at Ely Center of Contemporary Art, New Haven. He reviews the design and history of fountain pens on his YouTube channel "Drawing with Fountain Pens."

HUDSON RIVER ADVOCACY RESOURCES

Clearwater is a member-supported nonprofit organization whose mission is to protect the Hudson by inspiring lifelong stewardship of the river and its tributaries with innovative advocacy through educational programs.
www.clearwater.org

Forge Project is a Native-led organization that provides a model for Native cultural self-determination and leadership, fusing traditional and contemporary knowledge and practices to build community, public education, and collective action.
forgeproject.com

Hudson River Environmental Society is a nonprofit, non-advocacy organization that delivers the science behind Hudson Valley issues to citizens, scientists, and decision-makers by facilitating objective discussions, providing forums for rigorous science, connecting disparate views, and showcasing the region's natural heritage as well as facilitating and coordinating research in the physical, biological, and social sciences.
hres.org

Hudson River Foundation seeks to make science integral to decision-making about the river and its watershed and to support science-based stewardship of this extraordinary resource for all who live, work, and recreate there. The foundation connects the scientific community, policymakers, and the public with a wealth of information, providing materials ranging from research results and reports to educational opportunities to advance efforts to sustain the river.
www.hudsonriver.org

Hudson River Watershed Alliance is a collaborative network of engaged, informed, and active community groups, organizations, municipalities, agencies, and individuals that collaborate to ensure a healthy and resilient Hudson River watershed. It sponsors education, capacity-building, and networking events and provides people with tools, information, and resources to help protect the river.
hudsonwatershed.org/contact

Jamaica Bay Science and Resilience Institute is a partnership of the National Park Service, the City of New York, and the City University of New York acting on behalf of a consortium of seven other research institutions. Its mission is to produce integrated knowledge that increases biodiversity, well-being, and adaptive capacity in coastal communities and waters surrounding Jamaica Bay and New York City.
srijb.org

New York City Audubon is an independent nonprofit organization and grassroots community that works for the protection of wild birds and habitats in the five boroughs, improving the quality of life for all New Yorkers.
www.nycaudubon.org

New York State Department of Environmental Conservation conserves, improves, and protects New York's natural resources and environment and prevents, abates, and controls water, land, and air pollution to enhance the health, safety, and welfare of the people of the state.
dec.ny.gov

Department of Environmental Protection safeguards public health, critical quality of life issues, and the environment by supplying clean drinking water, collecting and treating wastewater, and reducing air, noise, and hazardous materials pollution.
www.nyc.gov/site/dep/water/nyc-waterways.page

Riverkeeper, founded in 1966, is a nonprofit environmental organization dedicated to the protection and restoration of the Hudson River from source to sea as well as the watersheds that provide New York City with drinking water.
www.riverkeeper.org/

Scenic Hudson is Hudson Valley's largest environmental nonprofit. Its mission is to bring people and organizations together to conserve rural and urban lands, create parks that connect people with nature, and protect the land, river, and communities at the heart of the Hudson Valley's well-being and vitality.
www.scenichudson.org

PHOTOGRAPHY CREDITS

The copyright holders and the sources of visual material other than the owners indicated in the captions are as follows. Every effort has been made to supply complete and correct credits; if there are errors or omissions, please contact the Miriam and Ira D. Wallach Art Gallery so that corrections can be made in any subsequent edition.

Fig. 2. © Alan Michelson
Fig. 5. © Reva Fuhrman; digital image courtesy New York State Museum
Figs. 8, 14. Photograph © New-York Historical Society
Fig. 11. Photograph © Christopher Pickerell
Fig. 17. Photograph by Andrew S. Bale
Fig. 23. Licensed by Munson Museum, Utica, NY/ Art Resource, NY
Fig. 26. © Shi Guorui
Fig. 27. Artwork © 2024 Estate of John Marin/Artists Rights Society (ARS), NY; image © The Metropolitan Museum of Art /Art Resource, NY
Fig. 33. © Lisa Sanditz
Fig. 37. Artwork © 2024 Heirs of Josephine N. Hopper/ Licensed by Artists Rights Society (ARS), NY; digital image © Whitney Museum of American Art/Licensed by Scala/Art Resource, NY
Figs. 38, 39. © Every Ocean Hughes (Emily Roysdon)
Fig. 41. © Estate of Joellyn Duesberry
Figs. 44, 45. © Anthony Papa
Fig. 48. © Center for Land Use Interpretation
Fig. 49. Artwork © David Hammons/Artists Rights Society (ARS), NY; digital image © Whitney Museum of American Art/Licensed by Scala/Art Resource, NY
Fig. 50. Artwork © 2024 Ruth Orkin Photo Archive/Artists Rights Society (ARS), NY; Digital image courtesy of The Jewish Museum, New York/Art Resource, NY
Figs. 51, 56, 57. © Athena LaTocha
Fig. 52. © Yvonne Jacquette
Figs. 54, 55. © 2024 Estate of Alvin Baltrop/Artists Rights Society (ARS), New York
Fig. 58. © 2024 Holt/Smithson Foundation/Licensed by Artists Rights Society (ARS), NY
Figs. 59, 60. © Jean-Marc Supervill Sovak
Fig. 61. © Courtney M. Leonard
Figs. 62, 63. © An-My Lê
Figs. 64, 65. © Marie Lorenz

This publication is issued in conjunction with the exhibition
Shifting Shorelines: Art, Industry, and Ecology
on view at the Miriam and Ira D. Wallach Art Gallery,
Columbia University in the City of New York,
October 5, 2024 to January 12, 2025.

The Wallach Art Gallery's exhibition programs are made possible with support from the Miriam and Ira D. Wallach Endowment Fund and our patrons. *Shifting Shorelines* was made possible in part by the generous support of the Terra Foundation for American Art, Wyeth Foundation for American Art, Dr. Lee MacCormick Edwards Charitable Foundation, Lunder Foundation—Peter and Paula Lunder Family, and an anonymous donation in memory of Stanley M. Blaugrund, M.D.

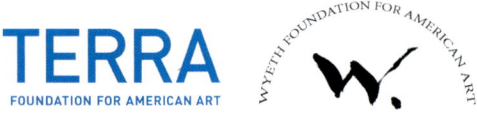

Copyright © 2024 by The Trustees of Columbia University in the City of New York.

All rights reserved. This book may not be reproduced in whole or in part, including illustrations in any form (beyond that copying permitted by Sections 107 and 108 of the U.S. Copyright Law and except by reviewers for the public press), without written permission from the publishers.

All rights reserved. No part of this publication may be reproduced, stored in retrieval systems, or transmitted in any form by any means, electronic, mechanical, photocopying, recording, or otherwise, without prior permission from the copyright holders.

Editors: Annette Blaugrund, Betti-Sue Hertz, Elizabeth Hutchinson, Dorothy M. Peteet
Managing Editor: Jeanette Silverthorne
Copyeditor: Pamela T. Barr
Research Assistants: Christine Yi-Ting Chen, Victoria Horrocks

Designed and set in Warkat and Referenz Grotesk by Laura Lindgren
Printed on 100 lb. silk text by Puritan, New Hampshire

Library of Congress Control Number: 2024940726

ISBN 978-1-884919-39-8